# A SOCIAL FAITH

BY Chester A. Qualls



Summary: Today many of us have computers, laptops, iPads and smart phones, and feel we waste too much time on Social Media. What if there was a way for your time online to count for the most important goal in the world... that is to show people how God lives through us, how he loves us and how he loves you too. This book is about that possibility and that opportunity.

ISBN: 978-0-9855211-1-0

# CONTENTS

# Thanks

Once I finally accepted the concept that I can't do it all by myself and that my best work includes contributions from others, I have embraced the concept. Fortunately I did not try to do this project completely on my own even though to begin I kept the number of people involved to a minimum. Of course my biggest help and support came from my wife, Kristin. She encouraged me and acted amazed that I could write a story. She not only provided emotional support but she went through the manuscript with me a couple of times during the editing process. She took what I wanted to say and helped me to make it even better, giving me ideas for what I could add to the fictional portions as well as the non-fiction parts. I would also like to thank my editor, Dona Watson. Dona went to the same college I did at the same time I did and it was nice to reconnect with her through Facebook and find out that she is a writer and editor. As I considered the concept for this book Dona provided encouragement, advice and tips. She often gave me links to read up on various aspects of the writing world. She has also been kind during the editing process providing many excellent suggestions. Dona edited in a way that preserved many of my original words while making me sound better than what I could accomplish on my own. She did also sneak in a few words of her own for which I am grateful. The truth is that if I take her advice I will come off as a better person than I really am. She has been a major help and I look forward to including her on future projects.

I also would like to thank my daughter, Ashley, for helping me with editing one of the stories and for providing the artwork for the cover of the book. I really appreciate that she has taken the time to do this even though she is busy with her college study load. I also very much appreciate the support I get from my parents and my wife's parents not pushing me to be somebody I am not and supporting us as a family with much encouragement.

I must also thank my friends on Facebook. You have all given me opportunities to practice my writing skills and also helped me to develop my thoughts better. I enjoy very much our discussions online and I hope we can continue them, at least until you agree with me.

# Foreword

Throughout this book I feel as if I may be poking too much fun at pastors. As I explain in the book, I got my degree in Biblical Studies from a Christian college. Today, many of my classmates are pastors. I respect them completely but since they were just my friends so many years ago and not leaders of a church, I may incorrectly feel comfortable enough to rib them a little. I should be more careful or they may include stories about me in their sermons.

So, as I say, I feel a little bad about disrespecting the position they hold in God's work on earth and at some points thought maybe I should take out some of the more sensitive things I mention about how pastors too can fall into temptation online, thinking that maybe I'm over-exaggerating a bit. Then as I was considering this I heard the news from a church I've visited about a pastor who had to step down from a wonderful church he founded, maybe without being able to come back, because he began an inappropriate relationship with a woman through Facebook. I don't know how far this relationship went. I can only hope that it never reached the physical world.

This news did, however, renew my belief in the words I say in this book. We must ALL be careful online. We must all take the steps to protect ourselves. We must pray, read the Bible and be in a community of people that can help us. Pastors must learn to become very open with their lives because as they move online, they will find that even they cannot hide. None of us can hide indefinitely online.

So I'm sorry pastors but you guys do need to take the same protections we do, or maybe even more than we do. I have heard Craig Groeschel of Lifechurch.tv talk about how he will not travel by himself. His wife usually has to stay behind to take care of the kids so Craig makes sure that a fellow male staff member comes along with him wherever he goes. This just makes sense to me to keep ourselves protected. We all need to do the same thing on the Internet as well. Stay in the light and be careful about back-channel communications. Learn to become completely open and have your own community of trusted people who watch over our shoulder so to speak. Don't let this keep you off the Internet but do make sure you are protecting your family and your ministry.

That said, I'm not just talking to pastors--we all have a ministry to protect. To some the Internet is a scary place but it doesn't have to be. It's a place you can be and you can have some great experiences that enhance your life.

My editor suggests I add another section and I believe she is absolutely correct. I need to warn you that some of the stories I tell can be a bit graphic. I have done that to shake people up a little and get my readers to think deeper about the subject of Faith expressed on the Internet. But I do recognize that some of you may be offended by this and may wish to keep young readers from being exposed to this material. That's understandable, in fact I'm often shocked by what Christian parents let their young children read, or watch for that matter. And which video games they allow. So please take this into account if you are planning to purchase this book for a younger reader. It may not be appropriate. I believe there are some things to which children are not yet ready to be exposed.

# MEET DARIAN

It had been a good Saturday evening for Darian. It started with a few drinks with friends, and then a cruise in his souped-up SUV to the clubs. As they passed a couple of girls on the boulevard, Jerry thumped the back of Darian's seat.

"Hey, is that Sacha?"

"I think so. Want me to pull over?"

"Of course, man!"

Darian rolled down the window. "What'cha doing ladies?"

"We're going to the club," Jerry called from the back seat. "Wanna come with us?"

Sacha jumped in through the back window and onto Jerry's lap, pressing her lips and goddess-like body against his rippling muscles, a sight that had become fairly common over the last few months.

"Anybody else want to come?"

One of the girls shook her head. "Nah. A friend of mine scored some blow and we're goin' over to hiz house to help him get rid of it."

"Suit yourself. Maybe next time!"

"For sure! Look's like Sacha's goin' with you guys though. Ha! As always."

"Enjoy the high! You'll be flyin' tonight!"

Everybody laughed as Darian pulled away from the curb.

The club's lights were down low and pounding music throbbed from the speakers of a great band that had been booked for the night. Scores of partiers filled the dance floor while others, like Jerry and Sacha, sat in a booth intensely enjoying each others' company.

Darian scanned the faces of the revelers, searching for ladies new to the club. With a smile, he spied two very attractive girls, apparently friends who, from the look of their newbie clothing, clearly didn't know their way around yet. At the least, they hadn't been to any of the local boutiques to pick up customized outfits. But that didn't matter to Darian. If things went well, he didn't plan on the clothing making much difference. Hopefully he would have them both out of it before the evening was over.

"Hello, ladies. I'm Darian. Haven't seen you girls here before, have I?"

"No, this is our first time," one of them said. "We're both new to town."

"I figured that. You have the NEW look in your eyes. It's a great place. There are a few dorks you should watch out for, but tell them you're Darian's & Jerry's friends and they'll back off." He gestured to the corner booth where his friend still hadn't come up for air. "That's Jerry over there with Sacha. They're an item--but not exclusively. Just having fun."

One of the girls grinned. "Yeah, we're here to have fun too."

"Well then, you found the right guy. Bartender," Darian gestured to the large man standing behind the bar. "Could you get me and the ladies another round?"

Like almost every weekend, Jerry and Sacha eventually toned things down to a snuggle and then a brief parting of the ways. There was always more time later in the evening to come back for more.

For one thing, Jerry liked to see if there was anybody worth fighting and, sure enough, tonight there was. He and a second macho guy took their fight into the alley behind the club, faced up against each other and the punches began to fly. Jerry ended up with a split lip, but with the brass knuckles he'd brought, he was pretty sure he'd broken the other guy's nose.

Meanwhile, Darian bought a few rounds for the two girls, but even after finishing a few drinks of his own, he didn't feel drunk at all.

"Well, it looks like my buddy has been in another fight." He gestured toward Jerry coming in, wiping his bloody lip. "He'll probably be wanting to leave so he can be...comforted." He jerked his head toward Sacha and winked at the girls. Jerry sauntered over to Sacha and buried his face in her silky hair.

"Yeah, I see what you mean," one of the girls responded with a chuckle. "Where ya goin'?

"I have a great apartment on the 32nd floor here in the city. My balcony has a stunning view and there's a hot tub big enough for all of us."

One of the girls flipped her glossy black hair over one shoulder. "Sounds awesome."

"I also have more drinks in the fridge. You ladies want to come over and we'll get to know each other better?"

The girls nodded. "Sure! Sounds good."

When Darian, Jerry, Sacha and the two new girls arrived at his apartment, Jerry and Sacha headed off to a guest bedroom, not to be heard from for the rest of the night. Darian brought a chilled bottle out to the girls standing on his balcony, admiring the view of the glittering bay below.

"This ice-cold vodka goes great with the warm water. You do like vodka, don't you?

"Sure do!" They settled down onto lounge chairs nearby. "We don't have any suits with us though."

"And I don't have any suits for you either. The soap's bad for the filter anyway so we always use the hot tub au natural. That's not a problem for you is it?"

The girls giggled in unison. "I guess not," one of them tittered.

Without their clothing, it was even more apparent that these two were newbies. Although challenging, it heightened the excitement for Darian as he showed both of them how things operated here. It seemed as if they had a good time as well. Darian knew he wouldn't remember the girls' names but maybe he would see them around another evening. He usually shared his hot tub with more experienced girls but two new ladies in the same evening made it worthwhile.

Eventually, the girls had to go but after they left, Darian sighed in contentment. It didn't get much better than this...

...and then it was time to sign out.

His real name is not Darian. That's only the name he uses for his avatar in Second Life, a virtual reality world on the Internet. Darian is a regular visitor and lives out many of his fantasies in Second Life. This night was rewarding but fairly tame compared to some evenings online. Nonetheless, he was going to have good dreams tonight.

Hey, no harm done, right?, he thought. None of it was real.

After logging out, Darian went to the darkened rooms of his two young children and kissed each good night on their foreheads. Then he climbed into bed, careful not to awaken his wife of over 10 years.

Yes, it was time to hit the sack in the real world because after a good night's sleep, he had to wake up, get the family cleaned up, have breakfast and get to church in time for the 11 a.m. service.

## MATTHEW 5:27-28 ESV

"You have heard that it was said, 'You shall not commit adultery.' But I say to you that everyone who looks at a woman with lustful intent has already committed adultery with her in his heart."

# CHAPTER 1

## BEGINNING

How did the story about Darian make you feel? Were you a little shocked? Did you wonder why Darian's wife doesn't know what's going on? Did you feel sorry for Darian's children? Maybe you are wondering how this guy can drop his kids off at Sunday School the next morning, sit in church and sing worship songs, and take communion. Maybe you are wondering why the church hasn't been a better influence on Darian.

Maybe you are Darian!

I don't know what this does to you. Could be that after you read Darian's story, you told the kids to get off the Internet. Maybe if you are a woman, you will look at your husband in a different way at dinner tonight. Will you?

For a while now I have been paying attention to what people say about their time on the Internet. I've had a few discussions and asked many questions because God has been working with me on the fact that many people don't seem to think of events that happen online as real, not like the real life they live in the physical world. Not true for me!

***Some Background***

Even before the Internet became so prevalent, I was a technical guy. I worked with corporate networks in which we had to consider or control a lot about what happens when people work together on a network. Now, of course, the Internet is one BIG network with everybody on it and I've been involved with that since the beginning. I love it. I love the way people work together and it's also been interesting to me how people interact online.

I admit that there have been temptations for me as well. Because of my background, I can hide online better than most. I know a little about where trails are left behind on a computer and how to get rid of them as well.

The first time I got onto Second Life I was extremely intrigued with the possibilities for this virtual world and I had heard a little about the debauchery going on and that intrigued me as well. So far I've tried to stay away from things that would influence me in this virtual world. I have, however, in my work as a computer support person stumbled across things that customers have done online, sometimes accidentally--and sometimes not. I've often had to deal with images that pop up randomly on a computer because my customer clicked on the wrong thing. They got a virus. Like I say, sometimes it was by accident but not always. Sometimes curiosity just wins out.

In my time on Second Life I've heard some interesting conversations between other avatars. I always see the best potential for the Internet and the many things a person can do online. In the real world I refer to myself as happily naive. I sometimes think I know what kind of evil things go on but then I will find out what a person close to me has done and I'm shocked. The same is probably true for me online--I am naive there as well. In each chapter I will share stories that come from my experiences and observations online, sometimes with a little bit of creativity but the truth is I'm not even getting close to what is really happening in many places online.

If we have been paying attention, we know there are terrible things happening online. There are bad people looking to hurt or corrupt our children. There are women (or sometimes men posing as women) trying to destroy marriages. There are images, conversations and information intended to move us away from God's plan for our lives. There are lots of these people. As I say, I've had to clean up computers that suffer from a mistaken click and it amazes me the number of beautiful young women who are willing to share their bodies in photos for anybody to see. There are a lot of people online for the wrong reasons. They are there 24/7, ready to leap on any opportunity.

Isn't it about time we had Christians online ready to leap as well... people ready to battle for souls?

***Merging My Worlds***

So I still remain optimistic for the potential that exists online. It excites me. As I read technical manuals, talk to other technical people, attend conferences and read blogs, I have also been reading my Bible, attending small group every week and going to church most

Sundays. Through the reading I have done and the sermons I have heard and the discussions we get into in small groups, I have been trying to merge my two worlds. The real world with the virtual world. I've never been very good at holding things back. I always try to be humorous, but I usually will ask or say whatever is on my mind. Many people wish I wouldn't. Some people are better at guarding themselves in public than I am. There again, this gives me something to think about regarding who I am in the real world and how I am in my own virtual world, the one in my mind and heart. So after my first few visits to Second Life and my first discussions on Facebook, I have decided to be ME wherever I am.

Many people are living two different lives, preferring to keep the two worlds separate. People such as Darian have decided that they aren't ready to be the same person online as they are in the real world. He doesn't want to be a father and husband who goes to church when he is on the computer. He has decided he wants to live a little, to be a party animal. Many people are doing this same thing, only in different ways and on different platforms.

Some people express their anger online, making inflammatory comments that they would never make in public or in front of real people. They feel free to do so, and actually they really are free to do that. Maybe it's good for them to work it out so long as they don't cross the line into criminal behavior but of course we hear stories about that as well. Online you have a choice just like in real life, you can become a mean hateful person or you can work out some of your bad character traits and actually become a better person in both places.

So when I think about the story of Darian as well as other stories that I come across online, the big question in my mind is "Where is the Church?"

### *Churches and the Internet*

Many churches today seem to look at the Internet and social networking sites the same way that most businesses do. They see it as a marketing tool, a way to advertise what the current sermon series is. The Internet also allows churches to share the awesome sermons that their pastors prepare each week by posting video on their websites. For anybody who's interested, they can check out the section: "What's going on at our church!" It's important to share the calendar, let people know there is an awesome children's ministry, youth ministry and maybe even help for substance abuse. And, of course, if you ever want to visit the church, the website must tell you how you can find it or how you can contact the staff.

Realize I'm not criticizing these things. They are very helpful and I do my fair share of checking church websites for this exact information. I work with a web team at my Washington church and these are all important topics we talk about. Team members spend a great deal of time looking at the websites of other local churches and also the big national churches who have over 20,000 people in attendance. Let's face it, most smaller churches don't have a big budget for technology so it makes sense to follow the practices larger churches have used successfully. Makes sense to me, and as I say, this is probably necessary and practical. While it does make many church websites somewhat similar, they have the advantage of being user-friendly (sort of), allowing visitors the ability to easily negotiate each site. However, the end result is that most church websites basically comprise a marketing effort.

Perhaps if a local church were more enlightened it might see the Internet not only as an advertising website but as a way to connect with people both inside and outside of their congregation, to begin conversations and to find out what really makes their people tick. One of the most current approaches used by churches lately is to set up a Page or Group on Facebook (not a profile which, at the time I'm writing this, limits friends to only 5,000).

### *Could We Have a Conversation?*

I try to keep up on numerous churches' Facebook pages and for the most part, I don't see many people interacting with each other as a community. Even less regularly do I see many pastoral staff engaging people online throughout the week. Rather one mostly sees pastors doing online what they also do in the real world on Sunday; they preach or read Bible verses. I don't suppose that is wrong and in fact, often these words can be just the inspiration we need during the week. But why don't more pastors take this opportunity to really engage people? Why don't they find out what makes their congregation tick? Why not ask questions more often? I have seen some examples of this but it's still somewhat rare. Why don't they interact more?

Probably one of the biggest reasons we don't see this (according to what I've experienced at churches I know personally) is that the pastoral staff is too busy to spend much time on Facebook or other social networks (at least they see it this way). They are in hospitals or ministry meetings or visiting people in their homes or in meetings. I have noticed that some are online more than others, but few often share their real life thoughts, feelings and work with people online. This is a missed opportunity for the pastoral staff to be there for a guy like Darian when he is about to enter his virtual world. If a church or pastor is participating in this activity, I salute you.

People DO need a website to find out the important details about the churches they attend. Yes, I did say "churches" because it seems that the days of the whole family attending only one church may be coming to an end, much like how one person no longer works for the same company their entire career. Because many families now attend more than one church during the week to meet the needs of their family, they need to keep up to date with each church. But on most church Facebook pages, I rarely see even the fans engage with each other. In the few instances where I've seen members interacting, they mostly enjoy posting comments regarding the content of the past week's sermon. However, more rarely do they share the joys or sorrows of their week with each other.

Some will see a church page as a place to spam about a new business they have started or to share their political viewpoints. This is why they need a moderator who actively checks the page and who is aware of and can defend the beliefs and vision of the church. The church also needs to have a policy about sharing political beliefs and business ventures on their page and state that policy in a prominent place. It's a nice thing for a church to brag about the businesses owned by their regular attendees in this new online community we have. The church should be doing what they can to help those local businesses succeed. But if a church has twenty realtors, they'll have to address that in their own way.

***It's Not Just Up to the Pastors***

So, as I say, these are some excellent ways for a church to have a presence online. Many do a great job and yes, I believe there are some missed opportunities. But that's what the local church is doing online for the most part. We do it because that's what everybody else is doing and, in truth, we need to look beyond the norm and find ways to serve people online.

However, when I wonder where the church is for Darian I don't mean local churches. I mean "The

Church" that is YOU and ME. Not only the pastors and the paid staff. Not only the volunteers at the local church. All of us, everybody who calls themselves a Christian. The Church seems to be missing out on a huge opportunity. In fact, trying to be available for all the hurting people online is so big that probably no single congregation can deal with or mobilize for this.

The Internet is an area I have come to think of as a missions opportunity. I didn't mention missions on church websites in my previous discussion but many churches post information about the missions efforts in which they participate. Whether these missions are local or in another country, they post what they are doing and often show pictures from teams that participated in a short-term mission trip. They may also show the missions overseas that they support financially.

But the place where I don't see much happening in the local church nor "The Church" is the mission field online. If you read the "Darian" story and understand what is happening, you will hopefully agree with me that we have work to do in reaching people online in social networks. As you read on, you will find at the beginning of each chapter, stories about other online communities where we could or maybe even should have influence as well.

### *God Has a Sense of Humor*

Missions!? To me, the fact that God has put on my heart anything that would be considered missions shows that he has a sense of humor. In truth, I would rather not think about these concepts as missions but in reality, I can't seem to think of it any other way.

What's the problem? You see, I have issues with missions in the traditional sense. I think it goes back to college and a class I took on missions. I didn't agree with things my professor taught me, which soured me to what I see going on in the field of missions and mission fields. A primary point of disagreement was the statement by my professor that "missions is a science." I was raised in a mathematician–college professor's home and was taught (perhaps partly in jest) that the only true science was mathematics. So to call missions a science was foreign to me and this made me begin to question other issues from the class as well. As a result, since then I have become more critical about what I see the church calls "missions."

I struggle when I see (as I have multiple times) two nearby churches, both with missions into a third world country or even to their own downtown area, and they don't work together to really make a difference. None of these individual churches can really apply the resources of someone like Oprah to help out in another country, as she has done in Africa for example. It's impressive what she has done isn't it? Working alone, most churches can't compete with Oprah and yet if they would combine their efforts, they could match or exceed what she has done and really make a difference. It would be impressive. Some of this happens but rarely and there are all kinds of political or protectionist reasons for why it doesn't.

Hopefully many pastors are reading this and catching a vision for the Online Mission Field and, yes, I am talking to you along with every other Christian. But I'm not just hoping for you to get your church thinking about a new ministry to witness online. There is some of that going on, which is a great thing. However, online witnessing from one or a couple of churches working independently is too limited for what is needed.

I am hoping to get just one church to start to see the Internet as a mission field. That one church is THE CHURCH. The whole body of Christ as the Bible talks about it. Every person who calls themselves a Christian--

that's who I want to reach with the idea that together, we may be able to share the love of Jesus with the whole world by using the computer that most of us have in our own houses. It is my hope to inspire every pastor and every Christian to stand up and say, "I am the Church and I care about sin which separates us all from God."

### *It's Not Just About Witnessing to Non-Christians*

On the Internet there is much work to do, far more than just online witnessing can address. Think about it--the fellow in my introductory story, Darian, was a church-going family man. He had heard the witness of God's good news and had even responded to it. Yet he let his curiosity get the better of him, which resulted in his sinning against God's good plan for his life and the marriage relationship with his wife.

The man in this story could easily be any man. All men have thoughts that, if not careful, could turn into an ongoing fantasy. It's important when this occurs for men to ask God for forgiveness. Unlike Darian's story men should not rehearse that fantasy over and over again with a live person on the other end. This reinforces it in the mind and I am of the impression that when rehearsed, such fantasies will become very real when given a real-world opportunity. I also know of just the one to provide those opportunities. He appears pretty early in Genesis soon after man and woman are created. Satan will be there to create the opportunity. As the scripture in Matthew notes at the beginning of this chapter, it's a sin between you and God long before it becomes reality. But, of course, after it becomes reality it becomes a sin that involves others and their relationship with God. So as I say, there is more than witnessing that needs to exist in online communities. There are relationships, friendships and groups that need to form to keep each other pure for our key relationship, which is with God. It's not a quick solution--it's a lifetime of involvement. This is a big commitment for The Church and many of the individuals in it.

***Internet Isn't All Evil***

Before we get too charged up about the dangers of the Internet, there are a few things to realize. First, not everything on the Internet is bad or evil. There are many very good things going on. We can find old friends from high school and college. For some of us it has been a while since we lost contact with them and now we find them on Facebook. I have been able to get together with an old college friend or even a recent friend from a different city when they posted on Facebook that they were in my town--or planning to be in my town. It has been great.

There are also many opportunities for new friends--as long as you are careful. You can get involved in discussions regarding just about any subject in which you have interest. You can look up facts and charts to assist in your discussions. I personally try to work on improving my personality skills online by learning to be less judgmental, becoming a good friend and treating people with respect. There are Bible studies and church services that you can attend online generally any hour of the day and don't forget those online sermons. If you are looking for answers you can find viewpoints and opinions that may give you what you need. There are all kinds of inspiration and beauty on the Internet. Tons of photos that will lift your spirits and videos that will give you the laugh you need for the day.

I wanted to mention the good things happening online because so many people I talk with hear about only the bad things that happen and for that reason they decide to stay far away from almost anything on the

Internet. They certainly don't want their children to go there but they also just can't see why they should spend any time online. This conflict is exactly what we will talk about in the next chapter.

***Are You Ready to Care for People Online?***

For now, what I hope you realize is that there is a mission field on the Internet. To whatever extent you have time to do so I hope you see yourself as a missionary or, if you are like me and that thought brings up negative connotations, then I simply hope that you will take the words of the Great Commission to heart and realize that God wants all of us to share the story of our faith and relationship with the whole world. He called us to share with our neighbors as well as those in the next town over, and then to share to the ends of the earth.

Hopefully I have you thinking that even though Christ didn't add "and the virtual world," you will consider this as a new mission field, maybe even the final mission field. There is a hurting world out there and many of the hurts can be found online for you and me to help. We can't all get to other parts of the world physically and there are countries, unlike America, where the masses don't have Internet access. But there are lots of people online, some like us and some very different. I personally can only connect with a few, but if we combined our efforts, deciding that we won't leave it up to pastors or professional ministers, then we could reach almost everyone that is online. And through those people we can hopefully reach the people who are not online. I don't have to be a pastor (and many people are glad that I am not), but I can stand up and say, "I AM THE CHURCH." And once I do that, I realize that I have a responsibility to share the story of my faith and relationship with everybody with whom I come in contact, even though it may not be the first thing I share with them as one would in street witnessing. To some it comes across more like salesmanship than about developing a true Christian and discipling them in their relationship with God. So to me, this isn't something we need to work up our courage to do, and go get in somebody's face. I really think it's as simple as trying to be their friend, and trying to spend some time with them.

If being a good friend and reaching people with the Good News we call the Gospel (which, after all, is what "Gospel" means) sounds like something you can do, then hopefully you will continue reading this book. If you just don't have time for people, this may not be for you. I don't believe you are like that, however, and I invite you to see the Internet as I see it.

For me, the Internet is an opportunity to meet new people and get to know them better. I also see it as a place where I can open up and let more people know who I am. I may not like everything about them and they certainly won't like everything about me but the Internet is a place for us to work all of that out. I do have to be careful where I go and what I share with strangers but I pray that God uses me on the Internet and that he convinces you to spend some quality time there as well. I hope you will begin to see that it's not a waste of your time to get on your computer and do things you enjoy doing. Are you ready for that? Are you at least a little open to seeing the Internet in a different way? I challenge you to keep reading—and keep your heart open to what God might have to say to you about this growing mission field.

# Meet Irene

"Irene is right, we should do something to reach out to the young ladies in our church." The grey-haired grandmother shifted in her seat and smoothed her skirt down with fingers bent from arthritis. Rose never missed the weekly Bible study with the other older ladies.

Her best friend, Myrtle, twisted her hankie in her hands. "But what could we do that would make them want to spend time with us? Any ideas?"

"Well, I don't think this generation would want to sit around and listen to hymns that were popular when we were their age," a third lady chimed in. "They would just think we don't understand anything about them."

"We could teach them how to cook!" Lois wore a wide grin and her eyes sparkled. Her cookies were known to be the best in all the church.

"That would be fun. Everybody has to eat," said another.

"But we would have to do that at one of our homes," Myrtle said, her brow creased with concern. "Their parents may not allow them over, and I remember when I was young I thought old people's houses were depressing."

They won't buy it if we try to act to cool, we can't very well do tattoos or body piercing.

"Well, what do they like to do?"

The ladies sat and looked at each other for a few moments, deep in thought, then all eyes fell back on Rose when she cleared her throat. "I know my grandkids spend a lot of time on the Internet and on video games."

Choruses of "yes" and "mine too!" came from several ladies.

"But it's such a waste of their day." Myrtle sighed and her eyes took on a far-away look. "When we were young," she said wistfully, "we got outside and played in the sun or swam in the stream near our house."

Yeah I wonder if they will look back and realize how much time they could have been doing other things, talking with friends. Learning things...

"My kids have tried to get me onto a computer," one little lady said timidly, as if embarrassed to admit it. "They bought me one but I just don't know what I to do with it past reading an email from them."

Even though Irene is seventy, she still wanted to feel as if she could do something for God and make an impact on the generation of her grandchildren. She and the other ladies agreed that they don't want to just sit around and grow old together. Just then, an idea came to mind.

"How about a craft?"

Myrtle tipped her head sideways and looked at her friend. "Like what?"

"It should be something that takes a while to finish, something that they could be proud of."

The ladies fell into thoughtful silence for a few more moments, then Rose caught her breath and sat up straight.

"How about quilting? We could help them make a quilt, it would take weeks to do and when they are finished they would have something to be proud of and useful on a cold night. Do you think they would like that?"

Irene nodded. "My granddaughter always admires my quilts. It would help them realize that things weren't always bought in stores."

"Well, I'm not very good at making quilts," Lois chuckled," but I could even learn something and I could bring cookies each time we meet with them."

Myrtle smiled. "Maybe they will even let us tell them what it used to be like when we were little girls."

"And maybe they will tell us what it's like today," Irene added.

All of the ladies liked the idea so much, they decided to make an appointment with the senior pastor to talk about this further and see if the church could advertise during the Sunday Service or at the youth meetings.

Meanwhile, Sandra, Irene's 15-year-old granddaughter is at the school library. She cast a quick glance nervously around the quiet room and wiped damp palms on the legs of her skinny jeans, then brought up Google and typed, "haven't had periods" and "pregnant". After pressing enter she found a women's health website where women can post questions and discuss their health issues. The particular discussion she found was actually a group of women discussing what it meant to no longer have periods after menopause. Sandra meant that she hadn't yet had a period because she was so young but, oh well, she signed herself up for the free account--using a false name, of course.

She posted: "I'm 15 years old and I had sex two days ago, today my stomach feels upset. I haven't ever had a period so can I be pregnant?"

The next comment in-line on the thread, a woman states that after going through menopause, you have to go twelve months without a period before you no longer have to worry about contraception.

Nobody answered Sandra's question. After all, this discussion was about menopause, not a 15-year-old worrying about being pregnant.

Three nights later, Sandra sat on her bed, a quilt that her Grandma made the only comfort she has as she cries. She still doesn't have an answer to her question and she feels like she doesn't have a friend in the world.

## LUKE 10:25-29 ESV

"And behold, a lawyer stood up to put him to the test, saying, 'Teacher, what shall I do to inherit eternal life?' He said to him, 'What is written in the Law? How do you read it?' And he answered, 'You shall love the Lord your God with all your heart and with all your soul and with all your strength and with all your mind, and your neighbor as yourself.' And he said to him, 'You have answered correctly; do this, and you will live.'
But he, desiring to justify himself, said to Jesus, 'And who is my neighbor?'"

# CHAPTER 2

## Too Busy To Play

First of all, I want to go on the record saying that I think the quilting idea was a good one. Maybe the young girls in the church need to spend time with these ladies and listen to their stories. By doing that they may avoid situations like the one Irene's granddaughter, Sandra, is in. She doesn't yet know if she is pregnant. The point is that there are all kinds of other ways for these ladies to help and minister to the young ladies if they are open to it.

We live our daily lives dealing largely with routine. Wake up, have breakfast, take a shower, exercise, do some work. Your routine may include other items or have a different order but don't most of us have regular routines? Very rarely do we concern ourselves with what is happening in another part of the city, another city or a different country. We don't have any interaction with those people because it takes a while to get there. We have to get in the car or on a boat or airplane, sometimes for hours, to go check in on what's happening. So often we don't. I'm guessing for many of us this is true.

But here's a question, do you care about any of those people? Do you care about the person around the globe, somebody in another state or on your own city block? You will probably say, "Yeah! I care!" So how do you show it? Do you give money to causes? Do you help your neighbor with yard work? What kind of interaction do you have with any of these people? Do you care enough to ask them what they need? Not as their political leader or a representative, but as a friend. Do you ask them? Get to know anything about them?

One of the best neighborhoods I ever lived in was a cul-de-sac and we all spent time in our front yards talking with each other. Often we would borrow tools or even help rake leaves or roto-till the garden (I had a roto-tiller). We even got together for some progressive dinners where we went from one house to the next enjoying a different course at each house. The kids would run from one house to the next while they were playing. I personally haven't had a similar experience in any other neighborhood I've lived in. It was a neat place.

I ask if you care and I'm sure you do. But I wonder how involved you are as well. I know it takes time to do that. I know I often don't care enough to do all that I should, but I want to find a way to extend my reach. I want a way to do more than I currently do. I can't afford to fly around the world and ask directly. Luckily I'm pretty involved with this thing called technology. Again, I really think the potential is awesome.

### *We Are All Connected*

Sometimes we get a glimpse of what is going on elsewhere by watching the news. A terrible tragedy in the Middle East, or China, Africa or New York. Maybe we get a glimpse into another part of town because the police have a fugitive pinned down and the news helicopters are capturing every minute. If you are anything like me, I watch these news stories with amazement but I don't really think there is much I can do about it. Probably because there isn't, I shouldn't try to get in the way of the people who can help, such as the police or fire department. Maybe that's just an excuse or a rationalization that I use so I don't have to get out of my comfortable house and get dirty. I just leave it up to others.

I have been involved with technology since the early days of the personal computer. I remember when IBM was the big name and Microsoft was just this little company that wrote an operating system for those computers, one of a few that existed at the time. I personally always loved networks. The big names for smaller networks back then were Novel and Unix. Microsoft didn't even have a server operating system back in those days. The reason I loved networks was that people were connected on them. These little low-cost workstations could be connected together and the ability of people to share and work together was exciting to me (I know, get a life right?). Today many of the computers are all on one big happy network we call the Internet. The reality of this is amazing and I don't believe we have yet seen what this connection is capable of doing.

***Buyer Beware***

Since I'm excited about the connections and the collaboration opportunities observed, I have participated as a user in the websites where the future is being made, just like many of you. My mind buzzes as I think about the potential and consider where this is all going. For many years now it has been my job to consult customers on their use of technology, to come up with a plan and implement it. So I talk with all kinds of people about technology to find out how they are using it and make suggestions on what they could be doing. I also recognize that there are a lot of people who are not as excited about it as I am. If they get a virus, they don't know how to fix it so they are scared. If they get online, they hear horror stories about what is going on and they decide to just stay away from it. Moms worry about what will happen when their children get online and some keep very strict rules about how it is used.  Personally, I have always involved my children with technology but since I work at home, I'm always paying attention to what they are doing so that I am fairly certain that they are not falling into the traps that do, in fact, exist online. You see, that's the problem. Your worst fears are true. There are bad people online who, whether or not they intend to, can trap you or your children. They can be a terrible influence on the Christian mind.

There are a number of pornographic websites online that cater to every twisted fetish that a person may have, and you can accidentally stumble into them. There are online groups not only for these sexual desires but also for strange political ideals or promoting outright hatred. There are places online that I wouldn't want my children to experience, or myself for that matter. There are places all around the Internet where evil is discussed and rehearsed. As I illustrated in the fictional story at the beginning of the first chapter, people are rehearsing the desires in their hearts regularly. Sometimes they rationalize it as being acceptable because, after all, it isn't real. Onlookers agree that it isn't real and therefore isn't worth worrying about.

Let's face it, in the real world I would not walk into a strip club with my Bible, nor would I walk into the headquarters of a local gang or drug dealer to preach at them. I would not want to happen upon a meeting of the KKK in the woods and try to setup a friendly chat. That's just dangerous. There are people who reach out to these groups of people and I admire what they do. Many churches just wait until these folks end up in prison and that's where they finally catch up to them to tell them about the love that God has for them. This works but wouldn't it be nice if we could catch these people early in their conversions to lust or hatred? Wouldn't it be nice if we could safely walk into the strip club and become friends with the young women there and try to be an influence on them?

In the "fake world" online, where people are developing these desires or practicing what is in their hearts, we have an opportunity to get to know people who we would never be able to reach in the real world. People that we are afraid to approach or maybe somebody who seems perfectly normal by outward appearance. Generally, these people can be approached and befriended without any worry of physical harm.

### *If You Are Still Breathing, God Still Has a Plan for Your Life!*

In the story at the beginning of this chapter, Irene and her friends want to do something to help the young girls in the church develop a relationship with Jesus and with them. Many of these saints of the church I have talked to never consider what they could be doing from the comfort of the computer they probably now have at home. It's fun to watch these seniors who never thought they would get a computer or have anything to do with the Internet become a part of some of the hippest things about the Internet. They may talk to their young grandchildren on Skype, allowing them to not only talk for free but also to do a video chat with them. This was something that was considered science fiction just a few years ago. Maybe they did it on Star Trek, but not in real life. Some of the really hip grandmothers are also getting their own Facebook accounts and are now connected with their grandkids, keeping up on their daily lives. Some, of course, are shocked at what they have found out.

I've talked with many seniors in the church and often they are the pillars that hold up even the coolest churches today. They not only pay their tithes but also donate time to do the little things that need done. I've heard more than a few express their desire to have an impact in the local community. They aren't finished making a contribution and many now have the time to do so. Irene could be seeking ways to be helpful online, not only to young girls, but maybe also to new mothers who are tired and worn out and feel as if they are at the end of their rope. They may be able to give advice that helps better than any parenting class and they may be able to keep marriages together. Sure there's only so much Irene can do and most likely she wouldn't have happened to find the post her granddaughter made, but maybe somebody else would have. With an online missions mentality, Irene can help the people she sees who need help and maybe somebody else will be in the right place at the right time to help Irene's granddaughter, Sandra.

***Who is My Neighbor?***

The scripture verse in this chapter, Luke 10:25-29, is a discourse that leads to Jesus telling the story of the Good Samaritan. While the priest and the Levite refused to help the injured victim of a terrible robbery, the outcast Samaritan had compassion on the man, bound his wounds, fed him and arranged for him to be cared for.

I'm not going to pretend to be a theologian of much accomplishment but I will say that this story, which answers the question, "Who is my neighbor?", is actually stating the rules of the New Covenant. Actually it's also summing up the Old Covenant (The Ten Commandments) and saying this is what God meant by giving us those rules. The young lawyer understood what the law meant, but he didn't get the practical application of it. Jesus, as he often does, tries to put it into a simple story that illustrates the application. When we come across those who are beaten up in whatever form or fashion that may take and lying by the side of the proverbial road, we are to do what we can to help.

Online there are many who have been beaten up by life. In their physical life they may be putting on a smiling face but online they admit to the damage that has been done. For some reason, there they feel safe and able to share. Are we Christians there to help? Or are we just trying to find a way to witness to people without taking care of their needs? As churches, we are seeing our church services put online for people to attend while in the comfort of their home, but are we listening to the people? Are we noticing that they are emotionally beaten up and bleeding? Sure, we can help the open ones who are willing to be honest about their needs but how about those who aren't so bold except when they are online?

There is bad stuff online and we could try to avoid that but the real people behind the negative and possibly criminal activity need good, Godly people to be available to talk to. Are we going to move to the other side of the road and just walk by like the priest and the Levite did in the parable of the Good Samaritan? Are we going to try to ignore the problem? Local churches have an important function to fill. They build physical structures and try to provide safe places for people to worship God and hear teaching about His Word and to bond with one another. That is the vital work of a local church.

However, the Bible also talks about a different kind of Church, one comprising all of us who believe. We don't occupy a building we are scattered everywhere. We need to display that our faith is more than just attending a celebration each Sunday and then walking away with the worship songs still in our heads. Increasingly, more of us are on our computers and some people, like me, are almost always on our computers. There is so much we need to do. Will we be there for a teenager like Sandra? Can we help her to find the answers she wants? Or will we ignore the pained and hurting we find online? The problem comes when we just say we are Christians but in our hearts we are not living the life that God desires. Who is the real me? Well, let's talk about that next.

# Meet Marci

Marco must be gay. That's what Marci told herself many times to justify what she had done. After all, what straight boy would make a joke about the Abercrombie & Fitch leather coat she bought for $295 plus tax with the last of her birthday money? He said that the collar--the Sherpa-lined collar--looked a lot like his poodle. He had completely humiliated her in front of her friends and Michael, a guy she was VERY interested in who went to her church youth group. Marci was only a Junior while he was a Senior and now she felt like he would never look at her like she had hoped when she bought the coat.

She could still replay the whole incident in her head.

Marci and her best friend, Carmen, were standing in the school's courtyard when Marco, Michael and a couple of other guys strolled over to join them.

Marco grinned her direction, a mischievous look gleaming in his eyes. "Hey Marci! New coat?"

Her girlfriends had just been admiring her coat and her heart beat just a little bit faster to think that the guys noticed too. "Yeah." She shoved her hands in the pockets, trying to appear nonchalant.

Marco nodded, his eyes running over the coat. "You got it at Abercrombie? I saw them there. The collar reminds me of my poodle, same color and all...but my dog looks better!"

Marci's eye darted to Michael and her heart fell to see him chuckle with the others. She could feel her face turn bright red as she excused herself and slunk away in humiliation.

Marci's humiliation turned to anger and, although she tried to hide it, her thoughts continued to smolder whenever she thought of Marco.

One day, Carmen came over after school to hang out with Marci. Carmen noticed Marci's coat hanging on her closet door.

"Why don't you ever wear your coat, Marci? It's really cool, you know."

Marci shrugged. "I don't know…"

Carmen plopped down on Marci's bed and folded her legs up underneath her. "It's because of what Marco said, isn't it?"

"Maybe," Marci replied, not wanting to admit his comments had gotten under her skin.

"He's such a jerk." Carmen frowned at her friend sympathetically.

Her compassion loosened Marci's tongue and she blurted out, "I don't even know what that was all about!" She paused a moment. "Maybe he's gay."

Carmen looked at her with eyebrows raised.

"I mean, what straight guy would pay attention to the women's section at A&F?"

Carmen chuckled. "Yeah, that makes sense."

Now that she had started talking, the words just poured out of Marci. "I wish I could get him back without anybody knowing. I don't want to blow up at him and have everyone thinking I'm the jerk."

"Hey, Marci," Carmen bit her lip thoughtfully, "I think I have an idea."

"What's that?"

"You could use Myspace. Nobody knows who is really behind the accounts. You could create a page for some fictitious guy, put up a fake photo of a hunk."

"Yeah?"

"Then you start making this guy friends with a bunch of people at school. You know, make sure you have cheerleaders, jocks, guys on the football team--people like that."

"OK, not following," Marci said although she was afraid she knew where Carmen was going with this.

"Then after you get this guy a bunch of friends, you change his sexual orientation to homosexual/gay and you have him post a list called "Guys I've Kissed." Then you put Marco's name on that list."

Marci gasped. "Carmen! That's cruel!"

"Yeah," she grinned wickedly. "But it would be funny. Just think...who's face would be RED then?"

"True," Marci conceded. "It certainly would embarrass him."

"Aaand, he would have no way of knowing it's you."

Somewhat reluctantly, Marci agreed, but then once they got started, she began to have fun with it, thinking of how Marco would endure the same humiliation he had dished out to her.

And it worked! In fact, once they added Marco's name to the list of gays, the news spread faster than they had expected. Some of the guys from the football team even beat up Marco and constantly teased him in PE. From time to time, Marci felt a little guilty, but then she'd think about Sodom and Gomorrah from her pastor's sermons and how much God detested homosexuality. Maybe that meant God would condone what she had done. Besides, it was all just for fun and it seemed to have put Marco in his place. Justice had been served.

At lunch one day, Carmen caught up to her friend in the hallway.

"Hey Marci, you hear about Marco getting beat up?" She kept her voice low so no one else could hear.

"Yeah," Marci grimaced. "I feel a little bad. He's had a black eye almost a week."

Carmen waited until a couple of kids passed by, then asked, "He doesn't know it was you, does he?"

"No," she shook her head. "I don't think he suspects anything. I say 'Hi' to him all the time and make sure I'm real nice." She giggled. "He does look a little silly smiling back--with the black eye and all."

"Well he got what he deserved!" Carmen huffed.

"I suppose so. But I still feel bad about the black eye."

"Really?"

Marci held up her pointer finger and thumb, showing a gap between the two. "Yeah. But just a little."

## Romans 7:15-20 ESV

"For I do not understand my own actions. For I do not do what I want, but I do the very thing I hate. Now if I do what I do not want, I agree with the law, that it is good. So now it is no longer I who do it, but sin that dwells within me. For I know that nothing good dwells in me, that is, in my flesh. For I have the desire to do what is right, but not the ability to carry it out. For I do not do the good I want, but the evil I do not want is what I keep on doing. Now if I do what I do not want, it is no longer I who do it, but sin that dwells within me."

# CHAPTER 3

## THE REAL ME

The Internet means different things to different people. It is a friendly place to some and a dangerous place to others. Through Facebook it's great to connect to new friends or find old friends. In fact, many waste a lot of time here unintentionally.

In the early days of the Internet (not all that long ago), you plugged a computer into a modem on the plain old telephone line and had to dial up a phone number to get connected. You would hear the numbers dial with different tones and then a tone would go out and then static and bong, bong, screech, screech, bing, bing, screech, screech, bong, the sounds of what was called "handshaking." Using this old and slow method, we were able to stay in touch with friends via email. Files could be shared in ways that seem cryptic by today's standards. There were also connections made between people in online Bulletin Board Systems and some of these developed into role-playing games. Usually in these games a person could pick a pseudonym and develop a new persona through which they could live out their fantasies. From the very beginning, people began to be someone online that they would never try to be in the real world.

I've heard people say that what happens on the Internet is make-believe. It's not the real world after all. These areas where people live out their fantasies are just that, not real, they think. I suppose one might say, "What happens on the Internet stays on the Internet." Well in this story of bullying it didn't stay on the Internet. We have heard of bullying that ended quite tragically. In our youth even as a Christian we can take the lessons of the Bible all wrong. I believe Marci in this story got it very wrong. One sinner is no different from another sinner... and we are ALL sinners.

### *The Oldest Profession*

During the early years of the United States, particularly out in the West where establishing a community was tough, young towns had to work very hard to survive. It took commitment and, of course, it took money. New commerce couldn't provide all the things that were greatly needed so the town council or city leaders had to come up with ways to pay for such services if they were to take advantage of potential growth.

Some towns held the promise of gold, while other towns became ports and outfitted those heading to their riches in gold. For some, their idea of making a living was to own a large piece of land. No matter what, people struggled.

More than once, a struggling town looked around to see what commerce was booming so that they could apply a tax to help with rebuilding costs or new infrastructure costs, and they often found a booming trade in the sex industry. Prostitution was known to be an abomination to God, but more than a few towns and God-fearing city councils decided that if you couldn't beat it, tax it. The sex industry allowed for the growth of many a young town. This has been true for the Internet as well.

The sad fact is that a lot of what we enjoy about the Internet is supported by revenue from the sex trade. In most Western towns, the sex trade is no longer legal but still it goes on. We no longer support our towns with tax revenue from those trades. In the Old West, preachers often came to town, rolled up their sleeves and began the work of sharing God's love. These pastors would not have made it if some in the town did not support them. Prostitution became illegal as a result. The church became a more important fixture in town than the brothel. Churches grew in size and in numbers.

This was no easy task and, actually, as large as some churches have become, it's still not an easy task today, even for them. Some estimates claim that no real progress has been made to the actual number of Christians in the last 20 years, particularly in the United States.[1] Even though we have seen large mega churches rise up, some feel their members are mostly pulled out of the community's smaller churches. Sometimes this happens merely by attracting young people as they go looking for their own relevant faith at a "cooler church."

### *It's Just Make-believe*

The Internet is not only a place where friends can meet but it has also been a safe place for fantasy to flourish. As a whole, we are a very creative people. God placed in us simple desires that help continue our species. He also decided to actually make it enjoyable for us. Thank you, God! He also placed in us desires for self-preservation and a desire to make ourselves comfortable. All of the ways God made us are meant to be good, but as I say, we are creative and there is another force at work on this planet--a force that wishes to separate us from the good that God created. Together with Satan, we have come up with ways to take what God created as good and turn it into evil. This is nothing new. It has been going on literally since time began. What is new is the Internet. We now have a new tool with which to work out our fantasies and this tool has a particularly interesting twist--it isn't real! At least many of us don't feel like it is real.

Leading up to the Internet, we have been working on this concept of "not being real" for years in many different avenues of entertainment. We watched movies in which a killer will hack up a body, and some might even share it with their young children, telling them,

"Don't worry, honey. It isn't real." We do this even in seemingly innocent movies with an emotional plot. Think about it, have you ever said that to your children? "It isn't real. It's just a movie." On the surface that's true--it isn't real. The Internet is similar--it isn't real either.

***"The Eye is the Gateway to the Soul." Herman Melville (1819-1891)***

While the Internet may not be real, it does have an impact. Few of us will watch a movie with murder in it and have a fantasy about committing murder. Some do, however. More people will watch a movie with sex in it and desire to live a similar life, but not all. However, just watching this in a movie can potentially move us to sin.

Now, I'm not going to try to put a big guilt trip on you. Rather, I'm going to admit a little secret to you...I AM a Sinner! Notice I didn't say WAS. I'm kind of taking the 12-step approach here and admitting that I have a problem with sin in my life. I will also admit right here and now that I can't control my sin on my own.

Phew...now that I have that out of the way, I want you to know that sometimes we sin by doing and sometimes we sin by thinking. So for some of us, it may not be a sin to watch a movie, even one with a fair amount of brief nudity or adult sexuality. But for others of us, it is and only God knows which one it is for you. Yep, he is aware of your heart. You aren't hiding a thing.

It's true that we can be innocently watching a movie, maybe even with a bowl of popcorn, and a scene pops up on the screen that for you could be a sin if you don't turn it off. Unfortunately, we can't control what the director, actors and editors put in the movie and it might have caught us unawares. It happens.

But now let's jump to the Internet. There we find websites like Second Life where we can interact in a number of places and even control the direction of the action. Just as in the story of Darian from the first chapter, he has control of what happens to his avatar. He interacts with actual people through their avatars and the actions he takes online often reflect his true desires and very real fantasies.

Second Life is not the only place where such things happen. World of Warcraft is a similar online game and wherever you get people, you will have similar behavior. As I mentioned previously, some old bulletin boards developed into role-playing games possibly modeled after Dungeons & Dragons, which may be the original structured role-playing game. There, people would develop a fake persona and would interact online with other people. The role-playing could involve any correlation to what actual people can do and also went beyond what is physically possible. These were all text-based games. People would type (in a conversational format) which choices they made or events that would happen. In addition, random events could be determined by a game moderator, who would also establish rules of conduct. When I say anything could happen, I do mean anything. As I said, we are very creative. Often to make certain events secret, a code word would be developed to indicate that behavior. So sexual activity online in those days may have comprised certain words or phrases that all the players understood but an outsider would not.

Today, youth do similar things with texting on their phones so their parents can't catch them. This is not real sex, of course, they reason. It's only pretend. And many believe it's just all in fun. Maybe in some cases that is true but, nonetheless, it is a fantasy which is being rehearsed over and over again.

Today we have avatars in a variety of online multi-player games that are actually 3-D characters that the players control. It is more lifelike than the old days of typing text on an online bulletin board, even though today the characters still interact by typing.

In Second Life and other online games, if a person has a microphone and speakers or a headset they can actually talk with the people around them in their real voices with no typing required. You can also purchase or create somewhat realistic sex organs, as well as actions that simulate sex. So today it actually LOOKS like sex, it's not just TALKING about sex or a secret code word. Even though people can use their own voices with a microphone, some people still say it's not real. Technically, they are correct. It is not real sex.

But I have to ask you, what is more real--the persona you portray when you meet people in real life or the thoughts you have that nobody else can see? That is, nobody can see them except God. Only God and you both know who you really are.

If you are anything like me, you can relate to the words of Paul in the book of Romans--sometimes I don't understand my own actions. I particularly don't understand my own thoughts. After all I want to do God's will for my life but no matter how I appear to the world, I know that I fall short of that goal often. I hate to say it but that is the real me, even if it's not what you perceive me to be. In reality, you are seeing the "me" I wish you to see. I am seeing the "you" that you allow me to see. But both you and I have other things going on and both of us, according to Scripture fall short of our own personal goals and we certainly fall short of God's will for our lives.

***Free to be ME!***

You may criticize the Internet for being what I have described it to be so far, but personally I love the Internet for this very same reason. You see, people hide behind masks they create in the real world, but online they are finally free to explore their real self. I maintain that this is the real them, their true nature coming out. In the actual world they can put up a wall, but on the Internet their philosophies and attitudes often become very transparent.

Thank God, I believe this is a wonderful thing! Are you confused now? Please don't be but join me in realizing that because of this transparency and because we finally get to see the real you and the real me, we have a wonderful opportunity to minister to each other. You and I can now work with the truth of our lives. We can say what we really want to. We have an opportunity to deal with these issues and actually improve. Notice I'm talking about me? When I first thought about this book I really wanted to help others, the many faceless people on the Internet. I also thought about you, knowing you would be reading this book and thinking about your own Internet habits and who you really are and thinking you could be better or do more. But I certainly wasn't thinking about ME. No not ME, after all I'm writing this book. God put it on my heart for YOU, right?

But as God so often does, he has caused circumstances to occur, has had me hear sermons and put books in my path that have broken me down until I realize that, indeed, I need this too. We need to not look at others as if they are the sinners and we need to witness to them. Whether we are on the Internet or at the Shopping Mall or eating at a restaurant or sitting in a bar, we are all on this earth together and we are all in the same condition--we all need the same solution. God seems to want us to hang out with each other and convince ourselves and each other to follow him. He may have even allowed us to create the Internet so that we could accomplish this task in a big way. He may have set it up so that we no longer have to pretend about who we are.

We are free to be who we really are and we are created to help each other.

In the physical world, we don't seem to be able to reach everybody and we don't seem to be able to affect all the people we do reach. Furthermore, we don't even seem to be able to alter the behavior of people who sit in our churches today. But you know what, like Paul, our hearts are often in the right place. We want to do good but just can't seem to make it happen.

### *Will the Real Me Please Stand Up?*

On the Internet, we can often hide our real world identity, but many do not hide their true identity and passions as we saw in the story about Marci. Which person is more real? Is it the Marci who walks around at school smiling at people or the Marci who created a fake ID online to ruin the life of a "friend"? Which person is more real in God's viewpoint?

I truly believe that God works through all the garbage we throw at him and provides a way for us to approach him. As we learn in Scripture, he already paid the price for our sin and gave us his Son as a special gift. Many of us who think about this fact are completely overwhelmed by it. It's odd to me that one of the ways

God heals this sinful nature in us is to put us near a bunch of other sinners, then by being honest with each other and living in each other's lives, we actually become better people. I also truly believe that the Internet is neither sanctified nor evil. It's not good or bad. It becomes both by what we do when we are on the Internet and what others do. It has brought people around the world closer together and allowed us to share the lives of people who do not live next door and who do not go to our church. The Internet has also given us an ability to connect with people who may even have different spiritual beliefs or political views than we do.

We are alive during a time when we have great opportunity. We do not have the luxury of ignoring the mission field that is on the Internet. As Christians, we also have an opportunity to do things better--maybe even better than we have done in the physical world.

***All We Need is Love...love, love, love***

I have tried to not come across as condescending or hateful, but I know some fellow believers have. Once we get out of our Christian circles, we find that we don't always have great reputations. Our actions precede us. I have even been blamed for some of the bad things that

Christians did during the Crusades or for actions taken in the name of Christians during the Holocaust. That's not fair but I'm not going to cry about it. After all, I accept that we don't always come across as representing God's love to other people. Was Marci presenting God's love to Marco? Would he think so?

I'm going to discuss this more, starting in the next chapter, but for now I just want you to see the Internet as something placed here by or at least allowed by God for us to reach into each other's lives and make a difference. Do you agree? Should I pray for you? Are you ready to be open about the real you and allow me to help you? And are you willing to be my friend and help me even if you discover the real me? Are you willing to join together and reach others? I know that's a lot of questions but that's the vision I have and that's what the rest of this book is about.

## Meet Tim

As a young boy, photography became a passion for Tim while working on his Photography badge in the boy's group at his church. The Commander of his group was an avid photographer and had shared his love of photography with all the boys. Tim also loved it.

When he was in high school, Tim took Spanish classes. He loved the sound and the literature of this language and continued Spanish courses in college while earning an engineering degree.

Over the course of several years, he sat through many a missions emphasis week at church, but one year, a presentation on Costa Rica sparked an interest in this beautiful, needy country.

Flash forward a few years and we find Tim, a bachelor and a successful engineer. While he enjoys his job, he loves to pursue his passion for photography in his off hours. At work, a friend had been telling him about a Google Group he had joined to connect with other people interested in his favorite hobby. That evening when Tim went home, he decided to see if he could do the same. He hoped he could discuss photography and practice typing in Spanish on a group of mostly Costa Rican photographers.

He sat down in front of his computer and found the website for Google Groups, then searched for "Costa Rica Photography," which brought up links to a few groups. While many of the groups only had one or two members, another had twenty-three members, so Tim clicked on the link "join this group." Later that night, he found that he had been accepted into the group. He looked on the boards and found a couple of interesting discussions, which he immediately jumped into by contributing insight and ideas and asking great questions. Before long, he had become a valuable resource to others and had also learned a lot in the process from several experts, which fueled his passion for photography to new levels. He even began to dream about one day making a trip to Costa Rica to take some of those beautiful landscape shots for himself.

As he got to know some of the photographers in the group, one of the members he hit it off with was Braulio. His new friend also loved photography and had taken some gorgeous landscape photos. But until a few years ago, Braulio had also taken sexual photos of children. By doing some research of his own and in his online discussions with Braulio, Tim found out that the creation, distribution and possession of child pornography had only become illegal in Costa Rica in 2007.

In one of their online discussions, Braulio told Tim, I could feel that things were changing a few years before they changed the laws. The police started arresting people in the businesses that were taking those pictures of children.

I did not want to be arrested and harassed so I started in 2005 to get out of taking those pictures. It took me a few months to complete some projects I had agreed to, but now I no longer take those photos.

Tim replied, I'm so glad that you stopped that business. You are such a great photographer, you shouldn't be doing pornography.

Braulio typed back, I stopped taking the photos of children but not the adults. I too love the photos of the landscape but for me it doesn't pay the bills.

Prostitution is illegal in Costa Rica but it is legal to be a "sex worker" and many of the girls make extra money by capturing their work in photos. Braulio is particularly proud of shots that combine the natural beauty of the Costa Rican coast along with the natural beauty of a Costa Rican "sex worker." Tim has wisely asked for Braulio to not share any of those photographs with him. Braulio says that he thinks some of the "sex workers" are still under the legal age, but he tries not to be involved with those he believes are. He tries to avoid being arrested by sticking to what is legal.

Tim is glad that his shock doesn't show over the Internet. He's also glad that he can sometimes just walk away from the computer and have time to think because the stories he has heard and the things that he discovers are quite disturbing and he needs a little time to think and to pray before he reacts. Fortunately, because Tim does spend time on his own spiritual needs, he has the strength to handle what he is hearing and learning.

Tim had gone looking for a chance to improve his Spanish, learn more about photography and about the beautiful country of Costa Rica but he never realized he would have opportunities to share his faith with a man involved in the pornography business as well as other members of the group. Thankfully, another one of the group members is a Christian as well, so he and Tim both have been praying for a chance to make a difference in Braulio's life. They believe that one day they will help Braulio to leave that business and hopefully at the same time will develop a closer relationship with God.

# MATTHEW 28:16-20 ESV

"Now the eleven disciples went to Galilee, to the mountain to which Jesus had directed them. And when they saw him they worshiped him, but some doubted. And Jesus came and said to them, 'All authority in heaven and on earth has been given to me. Go therefore and make disciples of all nations, baptizing them in the name of the Father and of the Son and of the Holy Spirit, teaching them to observe all that I have commanded you. And behold, I am with you always, to the end of the age.'"

# CHAPTER 4

## TO EARTH'S END

Clearly when God gives you a bad attitude about something you have to be careful. It's like he is setting you up. I feel like a horrible Christian at least one Sunday a year. That is the Sunday that many churches sponsor Missions Sunday. I feel terrible because as I am writing this, last Sunday was that once-a-year Mission Sunday. All the worship songs spoke of the Great Commission and the message was "just go." The church had not previously announced that this service would be focused on missions. As I sat on the first row realizing this was Missions Sunday and remembering last year's, I felt the life draining out of me.

I'm not proud of that and am possibly the only person who feels that way. Missions is supposed to be something to get excited about. After all, when I was a kid, I remember Missions Sunday as a typically boring service but at the end you got to go up front and look at the cool stuff the missionary brought from another country. There were items carved from bone, wood or maybe jade. There were brightly colored pieces of clothing and if it was a really good missionary, they had a weapon that a warrior had given them when he converted to Christianity. But just last Sunday, I was sitting there feeling depression coming on--then they played the same video from last year. It was a touching clip because it pointed out all the places in the world where there were no known representatives of Christ. And yet, there I sat feeling as if the blood was draining from my body. My wife knows all about me and my struggles with this, so she was not surprised when I said, "I have to go."

It is so wonderful to be in a modern-day church because hardly anybody noticed me leave--the house lights are usually dim with the spotlights focused on the stage, the TV screens show the words to the songs being sung, as well as close-ups of the guitar player, drummer, and singers, or in this case video clip. After my long walk out, I put a dollar in the coffee donations box and poured myself a cup. Returning to the lobby, I found a comfy chair and watched the rest of the service on the TV monitor. That allowed me the distance I needed to feel the life force come back. It also allowed me to be disconnected enough to take in some of what was said while I tried to figure out why I have such a bad attitude.

While contemplating this, I remembered that when I was in high school and college, it was popular for the young people to take a bunch of tracts and their Bibles downtown for street witnessing. I usually found a way to avoid going, feeling that it came off a little corny to the people walking by, as well as being an activity undertaken without much commitment by the participants. At least from me there wasn't much commitment. I felt that I might as well try to sell soap, something I probably would have done better anyway. For me, it didn't come across as being genuine. Did I really care about these people?

So what's the problem with missions, witnessing, and my attitude? After all, it's not as if I wish to go back to the original Greek and try to figure out a way to re-translate those passages in which Jesus tells us to "go." Turns out that's not a good idea. The Greek word for being God's witness in passages like Matthew 28:18-20 and Acts 1:8 seems to have a similar root as the word for martyr, so let's stick with the traditional translation.

As I contemplated this, I realized that I get frustrated with the way many churches approach witnessing and missions--primarily because their vision is so limited. I commend the many awesome people that have served in amazing ways around the globe. But if I conduct a search, I can find a huge number of ministries trying to reach select groups of people in the United States, Central America, South America, Africa and Asia. I can find more than one organization trying to reach the children in Honduras, or I can find many trying to deal with sex/slave trafficking. Why are so many small groups of amazing people working for the same goal but not combining their efforts? That is part of my frustration with Missions Sunday. I'm frustrated that we can't pull together multiple churches as well as Christians everywhere to solve problems around the world. If we are going to spend eternity together in Heaven celebrating God, then why don't we start working together now? Will there be a Baptist section in heaven? Or a non-denominational section? I doubt it.

### *Could the Church Eliminate the Need for Non-Church Charity Organizations?*

Let's consider hunger and health care. In the next few years, I expect to see some great things come from the Bill and Melinda Gates Foundation in these areas and more. Earlier I briefly mentioned the amazing things Oprah is doing to give back, but in comparison, the Gates Foundation is huge. Billions and billions of dollars of Bill Gates' fortune are targeted toward making a difference. If that isn't enough, Warren Buffet has thrown his finances into the foundation as well. This will total up to somewhere in the neighborhood of $100 billion, will likely be well managed and thus magnified in effect. A concept the Gates Foundation has is that some of the money used will actually earn a return, which can then be used again in other ways. This mix of charity and capitalism sounds pretty impressive, but I don't believe it would be if the entire Christian church would stand together to make a difference. The contributions of Bill & Melinda Gates Foundation would likely pale by comparison.

The best way to solve a problem is to be proactive, to address the problem before it even starts. The best book in the history of mankind that addresses being proactive in morality, in caring for our body, in being responsible for the earth that God put us on and for building up great marriages, families and finances is the Bible. I know that much of mankind has tried for centuries to distance itself from this book but the more we try, the worse off we become. We aren't that smart!

Our scientists learn all kinds of wonderful things about this planet and about our bodies. The more we learn about the complexity of creation, the more confused I believe we become. It's almost like a Tower of Babel–type of thing. We think we are brilliant, we put it in a test tube, light a fire under it and record the results, and along the way we begin to believe we know something. And we do, but the more we learn, the more languages seem to develop, the more complicated things become. Physicists and geologists both look at the earth and collect data, but they often disagree. The climatologist may tell us yet another story about the planet. As a result there just became more than one language about how to look at and talk about this world and our universe. Thinking we are smarter than God and talking about the same thing in different languages sounds a lot like Babel, don't you think?

I love science. It's a great thing, but the smarter we think we have become, the less we seem to think we need God. I believe nothing could be farther from the truth. We seem to rule out the idea that God is the source of all the knowledge in the world and that he is the reason there are atoms and calories and climate--that causes us many problems. Like I say, isn't that similar to what happened at Babel? OK, so a Bible Scholar among us may disagree with me that this is the message we should take away. I just believe that we have problems when we try to take God out of the picture. When we think science is the only answer to all of life's questions there are many answers that science has difficulty in answering. Then again, we can do the same thing with the Bible that we do with the results of our scientific research. We segment it, study it, translate it and pretty soon, we feel we don't need God any more to illuminate the wisdom contained in the Bible.

Please realize I'm not saying we should stop studying in scientific fields. After all, there are some technological mysteries that God did not include in the Bible. Fortunately, he left us something to do with our time on earth, since we were cast out of the Garden of Eden and the perfect life he created for us there. We need to explore this world and try to do the best job we can while we are here.

### *Basic Instructions Before Leaving Earth*

While I believe the Bible may not have the answer to safe nuclear fusion or the cure for cancer, it still has the important answers that can guide toward more productive and happy lives. God doesn't want us to be miserable--we achieve that all by ourselves. Rather, he gave us instructions for how to live. He told us what the meaning of life is. Isn't that the biggest question mankind has?

He also gave us excellent financial advice which, now that I've finally begun following it, has resulted in remarkable changes in my life, as well as decreasing the amount of stress I have about money, even while earning the least amount I've ever made. He has given us solid guidelines for managing relationships with spouses, children and friends. I don't know about you but relationships are important to me. They can make my week joyful or miserable.

The Bible also has simple advice for healthy eating and attitudes about food. One of the biggest problems we have with diets is not what we eat but our attitude about food. I can think of a delicious Italian dinner with mounds of pasta and I automatically want a loaf of bread fresh out of the oven with a nice crispy crust to go with it. Some extra virgin olive oil and balsamic vinegar would be great to dip the bread in. Then, if I start with the bread, I will no longer be hungry for the pasta with spaghetti sauce and Italian sausage but I will eat it anyway. Are you mentally sitting at the table with me right now? Why didn't we sit down with a large lettuce and vegetable salad with light dressing instead? We could then enjoy a cup of Spaghetti and be full. They are both spaghetti dinners.

I have had that large Italian feast I described above. I have tried to eat the veggies instead, but after years of trying to retrain my brain, the restaurant has more to do with my approach to this meal than the Bible does, when it talks about the sin of eating too much, eating far more than my body needs to live an active life. In fact, when I regularly eat the portion brought to me at the restaurant, I gain weight and can't maintain an active lifestyle. Yes, I know from experience.

***I Am the Church***

If we could work together to share these principles from the Bible, the church--everybody who would stand up and say "I am the Church"--would make a difference.

We would have more food to share, we would have more money to share and we would have a healthier lifestyle to share with the entire world. Reading the Bible and living its principles could eliminate hunger, relationship problems that lead to destructive behaviors, and twisted views of sex. We could also solve a majority of health-related issues and cut down consumption of products that lead to harmful effects on this planet. This alone would make the work for the Bill and Melinda Gates foundation much easier and maybe, if we took care of what was left, there would be no need for charitable foundations such as this. If we would work together to share what is in the Bible, the Church (that is, the Body of Christ) could be the most influential and life-changing force on this planet next to God.

So why isn't this happening? Instead, we seem to be running the other direction. There are many streets across the United States in which churches are lined up one side and down the other. The pastors may even gather every so often and pray together. But do they join forces to solve a problem? Typically not because they are worried that if they do, somebody from their congregation will see how cool it is at the other church and start attending there.

Why don't all those churches join together, buy a piece of land across the street and build the most awesome water park in the city? A place where all the families would want to bring their children regularly. If churches worked together we could accomplish more. That is true in the real world and on the Internet.

And yet in the case of the water park, I don't believe it's going to happen. One problem is that new churches are popping up around the country and attracting young people from existing churches to join together and build new massive churches with light shows and video effects and cool preachers who wear torn jeans rather than a suit or a robe while they preach. While I like those churches too, I'm wondering if we're reaching new people. Sometimes we do, but it seems like we could do better. In addition, are we even ministering effectively to the people who already attend our churches? How do we find out if they have a need if they don't tell us? If they don't fill out a prayer request card or form online, then how do we know to pray for them? How can we meet them when they are questioning their faith or their marriage vows? If we try to add all those responsibilities to the burden of the local pastor, we may be unrealistic in what we think they can handle. While I know they are directed by God and are often in the right place at the right time, I believe that God also expects a little something from those of us who are not part of the paid staff in our local church.

Remember the story from the first chapter? We met Darian, who took his family to church every week. His real character needed help. We talked about the real you and the real me in the last chapter. I have read statistics stating that in the church, there are no fewer divorces than in the general non-church population. In fact, I have heard there are more.[2] It seems like we should be able to claim that Christians experience fewer divorces than those outside the church, but we can't. I have heard that the number of youth in the church who have pre-marital sex equals roughly the same percentage as the local high school. The same number of kids are taking illegal drugs as well. I have even read that a shocking percentage of pastors regularly view pornography online. There is a lot of work to do in our churches and, in some cases, with the pastors of those churches. We have an opportunity to reach many of these people online. But even if no pastors needed this ministry themselves and wanted to minister to the hurts of people on the Internet, they clearly can't do it alone.

## *Is This a New Sermon Series?*

I believe this is not just a message for pastors so they will begin a new sermon series, I hope to inspire the entire church to join together on the Internet to make a difference. It doesn't matter if you are a Catholic and I am a Pentecostal, non-denominational or some other branch of Christianity. It only matters that I want to reach people with the good news that is in the Bible. It doesn't even matter if you are perfect or not. I certainly am not--if you ask my closest friends or people I went to college with, they will tell you that I do indeed have some character flaws.

It matters if you want to be the real you everywhere and particularly whether you understand that God's opinion of you does involve the real you and, by the way, he loves you. It's your real self that God looks at, not the mask you wear to church.

## *Making Disciples in All Nations*

The monumental problem we have is the Great Commission found in Matthew 28; this is where, after his Resurrection, Jesus tells us to go to "make disciples of all nations." But if we don't work together, how are we ever going to accomplish that task? I think the Internet has provided us the perfect solution. Not only are people finally showcasing exactly who they are on the Internet, but now we can also use it as a tool to reach people all over the planet.

I know that you may not think what happens online is real, but there are real world connections made every day because of virtual connections going on. Do you know somebody who found their spouse online? I don't think that's a terrible way to go, but if you are looking for sex online, I would say, yes, it is terrible. But there are many people legitimately trying to find a spouse; perhaps they already had a bad relationship founded through traditional dating so now they are a little more committed to telling the truth about who they really are when they are online. What good does it do to lie if you want to spend your life with another person? In some cases, online services can provide a superior solution.

In the real world, you might meet somebody to whom you are instantly attracted to. Because of that physical attraction you may be less than truthful about yourself in order to spend more time with this marvelous creation of God. Hopefully, you are not extremely untruthful because if you are, the relationship may not last long. Now, I know there are people who lie online in order to make connections that will lead to physical relationships--just like in the real world. But when somebody is looking for more than a purely physical relationship, it doesn't pay to lie.

This is probably the most intimate of all connections that can be made online. Not all connections have such serious consequences. There are many other wonderful, beneficial connections that can be made on the Internet--as well as horror stories. People with similar interests are starting groups all the time on the Internet, just like in the story at the beginning of this chapter with Braulio and Tim, and these associations can eventually lead to real world connections.

There are also negative, hateful, mean-spirited groups that meet online. This is a tricky area but can also provide an opportunity for us. In the real world, a guy may not admit he is a racist but he may be participating online with a group of others who spur each other on. It can be difficult to deal with such people, but at the same time, it can be a huge opportunity to reach out to people with a more positive message.

### *6 Degrees of Separation*

Not every person on the planet uses the Internet, particularly in select, more disadvantaged countries. However, many people are and these are people we can reach. In fact, I can reach people online who live closer to those who do not have access to the Internet--they may be able to walk in and out of villages where I would get killed. This is powerful. We can befriend someone online and, if we are being real and God's love shows through us, we can reach this person and possibly even some of his friends. We can form small groups online to interact with these new Christians in other countries--we can disciple them and as they practice their faith and learn how to share, they will be able to reach the nearby village that does not have an Internet connection. In the past it was difficult to get Bibles into some countries but online we have many options. Even if the Bible websites are blocked then scripture can be shared in many other ways.

When this happens, we have just done the greatest thing possible for that country. Rather than come in as the all-knowing and powerful westerners, we have developed a local team with the heart to reach those around them. Their combined talents and resources can be put into play to reach those nearby who are in need. If it becomes necessary, we may even be able to direct some of our vast resources to help out. I do emphasize "IF" because sometimes we westerners do damage when we apply our logic to other countries and cultures. We need to make sure that what comes into their country will truly have a beneficial impact and we need to not be quick to flood them with attempted duplication of our pampered lifestyles. There is great opportunity, however, and if we have properly developed a group of local people, we can make a difference from a long distance away. With the Internet, it doesn't matter where people are. Our new Internet friends from another country may even show us up and somehow reach out and become a friend of my next door neighbor whom I have failed to connect with. While it would be sad that I have never been able—or tried—to connect with them, there would still be rejoicing in heaven if this actually happened.

It is possible for us to be a powerful force worldwide and I expect that every friendship we develop could actually end up benefiting us--you always receive from giving. In the second chapter, we met Irene. Irene may not be able to connect with her own granddaughter for many reasons, none having to do with her failing. The great news is that some other grandmother may be able to help through the Internet and do so simply by paying attention and listening. We have been given an awesome responsibility to share the greatest gift ever. We now have technology that makes fulfilling that responsibility--and sharing that gift--an amazing possibility.

As I said before, we can't leave this task up to pastors as they generally don't have enough time to make it succeed by themselves. Rather, the tools have been created and shared with many who are not even all that technical in nature. Almost everybody can use these tools and once they do, they can then share their other talents online. The church can now literally be everywhere.

### *A Vision for the Future*

I have a vision of the Internet bringing all Christians together to accomplish something that we never could in the real world. I hope to get "real world" groups of Christians joining together to discuss the Bible and to perform acts of service for our fellow mankind. In an effort to do this I will begin an organization called iMChurch. A Facebook page will be the beginning place for people to connect and hopefully form groups.[3] These iMChurch groups will recognize all believers as part of the Body of Christ. Hopefully by the time you read this book, these groups will be starting to form, but just like churches that have started in people's living rooms, I recognize that this online "small group" will begin in the virtual living room of my house. I hope to find people from numerous churches around the world who want to make a difference online, people who want to be real Christians and who want to find others who are being real online and not living Christian lives. I do not hope to replace the local church. I personally love the variety of viewpoints and opinions we have in different local Christian churches. We need different types, and styles of churches because people are different and we all need a community where we feel comfortable. I hope for the iMChurch organization to be a support to the local church. It's about helping people find that local community where they fit but also coming together from different local churches to show God's love by meeting together and performing acts of service for others in need. We are all part of the body of Christ and we need to not figuratively cut off an arm, we can actually be part of a church right next door as well as join with The Church around the globe. That's the beauty of the Internet.

I'm not sure where iMChurch can go from there, but I have some ideas such as meeting in small groups, in person and online, for spiritual support and joining together to help the needs we identify. I know, however, that all my ideas will pale in comparison to what can happen if more people get involved and God shows up. Maybe we can power up to be an energy source in both the local church and around the planet.

Are you ready? Hopefully I've made my case. But before we jump in and begin our online mission, I think we need to prepare a little bit. You don't have to go back to school and get a degree in missions but you should still make sure you are doing the basics. What are the basics? That's what the next couple of chapters are about.

## MEET JOHN

Many years ago, John would never have imagined he would be sitting down at his laptop for a prayer meeting. When he was in the mission field, personal computers didn't exist. As a missionary, John had spent many years with his wife and family in first one country and then another, meeting people, starting churches and, of course, prayer groups. Spain had been his last assignment but that had been years ago and now, as a retired missionary, John lives in the United States. But today, even though he was about to use unfamiliar technology, he was also about to do something very familiar. John would be meeting with certain young people who wanted their work to count for

God and would be talking to them and praying with them for that work and their lives--just like he used to do in the field.

Today John would be praying with four young people: Eugene, Mary and Stephen (a married couple), and Santiago. It all had started when John's nephew Erick introduced him to a new couple in church.

"Uncle John, this is Mary and Stephen. They are moving to Honduras to work with children." Erick turned to his friends. "This is my uncle, John. He has been a missionary to Spain as well as a few countries in Central and South America."

As John shook the young couple's hands, he noted the bright sparkle in their eyes. It reminded him of himself when he was preparing for his own first missions trip abroad. "Nice to meet you," he said. "What kind of work will you be doing with the children?"

"We are both teachers," Stephen said, his grip firm and confident. "We plan to start a school for children who are not getting an education, and also to feed them because they don't get regular, healthy meals."

John felt a smile cross his face and he shot up a prayer of thanks, knowing how desperately this young people were needed in Honduras. "I will pray that God will bless you!" He meant it with all his heart.

"We greatly appreciate that," Stephen said, Mary nodding in agreement. "In fact, we know you have a lot of experience in Spanish-speaking countries and we were hoping that we could talk with you regularly."

John nodded. "I'm sure we could work something out."

Stephen asked John if he knew about Skype, to which he simply shrugged. "I've heard of it. Does that count?"

Stephen chuckled. "Well, if you have time, maybe we can get you set up so that we can get together with you to talk and pray often. Do you think that might work?"

"I can help," Erick interjected excitedly. "I've been trying to talk Uncle John into it anyway."

John grinned sideways at his nephew. "Yes, that's true. He has. I don't know much about this computer stuff but it would be an honor to pray with you and be any help that I can."

After John and the young missionaries traded contact information, Erick promised to stop by his uncle's house in a few days. As promised, he got a system in place and showed John how to use it. Soon, John was in business and before long, the group began to grow.

John came in contact with old friends he had originally met while ministering in Spain and soon their son, Eugene, was added to the group. Eugene is in Spain and is a software programmer for a company that builds car parts. John knew Eugene's family from the days when he was ministering in Spain, his parents had been children at a church that John and his wife helped to start. Eugene then asked if one of his Facebook friends, Santiago, could join them. He had recently moved to Buenos Aires to attend the University and learn to be a veterinarian, but was feeling very alone in his new home.

John readily agreed and soon he had opened a Facebook account and they were all on each other's friend lists. And now this group had been meeting to pray for more than five months, meetings that had become a highlight of John's week. It was such a blessing to be able to minister to these wonderful young people who were living for God--and in Spanish, a language and culture he had come to love and sorely missed since he had been back in the States.

The added blessing was that John could sit in his comfortable chair at home in front of his computer, login into Facebook and then into a new website designed for video conferences. John shook his head. When his nephew had first showed him how to run the programs, he didn't think he would ever understand but now he feels like a pro. He begins the video conference by logging into the website with his Facebook account. Once he does, the blue light next to the built-in web-cam on his laptop begins to glow and one of the squares on the website shows his face, this is his video feed. Next he sees that his friends are already online because their pictures have a little green dot rather than a grey one. After a few clicks, the entire prayer group was connected, each one sitting in a comfy chair at home. Onscreen, John can see the faces of all his new friends. Now that they are all in the conference video call, they begin with greetings and sharing a little how each one's week was going so far.

Mary related how Manuel, a little boy at their school, had been bitten by a rabid dog and was in the hospital for treatments his parents couldn't afford. Stephen added that they needed to find funds to help the family and, in addition, they were seriously considering whether to find a way to put a fence up around the school to protect the children from the wild animals.

John thought of a friend he had made on the mission field. "I have a contact in your area that might be able to help with the fence."

"And I can ask my church if they can help donate money to help the little boy," Eugene added.

Wide-eyed, Mary and Stephen looked at each other, then back at their friends. "Praise the Lord," Stephen said. "He definitely brought us together for a reason."

John nodded in agreement. "It's an honor to be able to help." He reflected briefly on how he had come to treasure these four people and their friendship, then it was time to get down to business. "OK, let's pray. Lord we come before you today to ask for you to help little Manuel..."

After they had finished and each one had gone back to their lives, John turned off the computer and sat for a moment, just staring at the black screen.

"Thank you, Lord, that you have made this technology available to me, technology that has allowed me to still be involved in the cultures and work I love so much, even though I can't be there myself." What a blessing to be retired and still able to serve--and in three countries at once!

# Colossians 4:2-6 ESV

"Continue steadfastly in prayer, being watchful in it with thanksgiving. At the same time, pray also for us, that God may open to us a door for the word, to declare the mystery of Christ, on account of which I am in prison--that I may make it clear, which is how I ought to speak. Walk in wisdom toward outsiders, making the best use of the time. Let your speech always be gracious, seasoned with salt, so that you may know how you ought to answer each person."

# CHAPTER 5

## A GOOD START

In the last few chapters I have been making a case that there is work for God's church on the Internet. I have hopefully convinced you of this. I want to get this book into the hands of more than just pastors and more than just technology-minded people. You don't have to be a technical whiz or a pastor to begin a ministry on the Internet. In fact, by the end of this book, I hope you will think of your time on the Internet as something far greater than just a ministry.

I believe that there is much more we can do than just preach at or witness at people online. I say "preach at" and "witness at" because often this is how people feel. It often seems like we are pointing a finger in their face rather than truly sharing good news with them. So if you are interested in moving forward, I believe I can offer a great place to start your new online attitude.

### *A Great Start is By Praying!*

First of all, I'm not trying to insinuate that you're not currently praying. I'm also not going to try and pretend to be some kind of perfect prayer warrior because I'm not. Let's just say I could definitely spend a little more time with my head bowed talking with God and hearing from him. But I have learned over the years that prayer helps. I believe in prayer. I have also had my faith shaken regarding prayer.

### *A Lesson in Prayer*

The first car I ever had needed a new stereo and speakers. I loved turning up my music nice and loud and that required a better stereo than the one that came with the car. I didn't have a lot of money but being the spoiled kid that I was, I didn't have to spend it on gas, so I bought a stereo with a cassette tape player and had it professionally installed. Now I had something that I had spent my own hard-earned money on. I didn't want it to be stolen so I prayed that God would keep my radio and tape player safe in my car. But I also prayed that if God did wish to have somebody steal my new stereo that he would only allow somebody to take my stereo and tapes if they would listen to them (mostly Christian music) and get saved in the process. I figured that covered me and if God did allow my stereo unit to be stolen, then I had the comfort of knowing that God would use it for a good purpose. I never had that stereo unit stolen from my car, nor did I have any of my tapes stolen. That prayer seemed to have worked.

Fast forward a few years (my tape player didn't have skip) and I had a somewhat successful computer consulting business and a new family home. I grew accustomed to having my prayer time while walking on my treadmill, thanking God for the many blessings he had provided for me, the great customers, a wonderful family and this great house. I also prayed that God would continue to bless me. I still have the wonderful family and the treadmill.

Unfortunately, though, I had to witness the rest crash and burn despite my prayers to the contrary. I couldn't figure out why those prayers weren't answered. In retrospect, I wish that God had allowed my stereo to be stolen and answered my later prayers to preserve my business.

Now, however, I do know a little bit about why God answered my prayers as he did. It turned out God had a lot to teach me over the next few years, and I am finally starting to understand why he needed to get my attention. I actually did need all the distractions removed so that I would focus on what was important. I thought I was paying attention--I went to church, I taught Sunday School, served on the church's Board of Directors, and more. But I was still missing some of the simple but vital things of life.

***Prayer Warriors***

As you can see, I'm not an expert on the subject of prayer. I don't have a great Bible study on prayer in this chapter. I simply suggest that you look at some of the prayers in the Bible, study them and figure out which ones God liked. Look, of course, at Jesus' prayers. You can see one of Paul's a few pages above in the Biblical perspective. You can go back to the Old Testament and look at David's prayers (and the prayers he turned into top 40 hits on a regular basis... Psalms). Also look at Moses' prayers. That is all the Bible study on prayer that you need. I was just taking a quick break and telling my wife that I was writing about prayer then as often happens while talking to my wife I realize something. I didn't pray for God's guidance before starting this chapter. That alone should secure the fact that I will never be asked to teach on prayer at my church. That's a little of a relief.

***Without Ceasing***

What I am going to do is suggest a few things to pray about regarding the subject of this book. After all, even if you don't feel you have a place in ministering to people online, you can pray for those of us who do. I don't have to go to Central America to educate children but I can pray for them.

You can start by praying for the very real people on the Internet. Thank God that people are being transparent online and letting out all their sinful desires. Why? Because now we can finally see them for who they really are and we can work with that. Pray for all these people. Pray for the pastors who are viewing pornography online. Pray for the young teen who enjoys virtual-killing of people too much and fantasizes doing it in the real world. Pray for the girls who are giving up the beautiful sexuality that God has given to them for all to see or read or listen to online. Pray for the family man who is rehearsing his fantasies online through an avatar in a massively multiplayer online (MMO) game, virtual world, in other chat rooms, or online groups.

Of course, we should also pray for people who are trying to share their faith online. We can pray that they will have opportunities to do just that. We can also pray that they will stay away from temptation because we are all sinners. Let's thank God for all the people who are online. Some have Christian backgrounds and some are in parts of the world where Christianity is not popular or may even be outlawed. Thank God for the different viewpoints that are represented in every topic we can think of. Thank him for the various political viewpoints online. Thank God for those who support the President and those who don't. Thank God for the people online who are not in the United States and if you are in another country, thank God for those who are in the United States. Thank God for the many religious groups and differing viewpoints on scripture which are portrayed in blogs and in various discussion groups online.

Next we can pray that the safety people feel online will begin to crack while they are exposing their real personality, that walls people put up around themselves in the real world will not only begin to crack but will begin to crumble. Pray that the real person they are exposing online will begin to show through in the real world and let's pray that they can find help before it's too late and their destructive behavior spreads into even more lives. They are now rehearsing it and let's pray they slip up. Cruel I know!

Pray also that people will begin praying with those they have met online or in their small groups at church. Share these prayers with each other. Why not start to form online prayer groups as well? All of the tools that are being used to bring people together for many reasons can also be used to get people together to pray.

Pray that people will begin to realize that what happens on the Internet doesn't just stay on the Internet and that it isn't as fake as they had hoped. Help people to see their own real selves, to realize that God sees what is in their heart. God knows everything about them, all their evil thoughts and sinful ways and yet he still loves them. So let's pray that more people understand that--people who are on the fringe of their faith as well as those who are actively involved in their faith. Let's pray for people whose faith in God has been shaken because they don't understand why he did what he did in their lives, or maybe why he didn't do that which they were sure a loving God would do.

We should also pray for ourselves. We all have things we hide and walls we put up to protect ourselves or to hide our faults. We could pray that God will deal with these issues in us (ever so gently, please Lord). Truth is we should probably pray for ourselves at least as much as we pray for those other people. I don't know, maybe we should go back and look at those prayers from people of faith in the Bible. Do they pray for themselves? Do they ask others to pray for them? I bet they do quite a bit. Look it up and let me know.[4]

While we are praying for ourselves, let's ask God if he has a place for us to share our faith online. Ask God to reveal if you should get involved anywhere. You may not feel comfortable stepping into an adult section of Second Life or making a comment in a women's health forum. I'm pretty sure God isn't going to call me to answer questions on a women's health forum. Not every place is somewhere you or I need to go. There are many places I wouldn't be comfortable online, but you may be able to go places I can‘t. I would also not have much to offer in an online photography club but you might.

I can join groups related to technology or network administration. I can provide answers on those subjects for forums regarding technology. Maybe I can make some connections with people there. I also very much enjoy discussing politics and business online. I love to get involved in discussions about the balance between capitalism and socialism. My wife thinks I'm crazy because these discussions stress her out, so obviously that's not a place for her. I, however, love to learn from people with different viewpoints than I.

### *Where Do I Fit In?*

You have a place online no matter what your hobbies, skills or passions are. There is a place even if you are not very technical. You can connect with people and begin your new virtual friendships. At first, they may not seem rewarding to you but that is also something you can pray about. You can pray that God shows you how you fit into the online world. And remember, when I say "world" I do mean the entire world.

You never know where you are going to connect with somebody or who they really are. You may think you are connecting with a young woman when you are actually connecting with a middle-aged man. Who knows why they have picked the persona of a young woman but, regardless, there is a possibility that God can use you in that person's life--you may get a chance to share God's love directly or you may have another online friend who will be able to connect. Two of your friends who didn't previously know each other may connect in a meaningful way. Only God knows what he has planned for us in using the Internet for his kingdom. I just pray that I will be in his will with the connections I make online. I pray that you will be in God's will as well.

Prayer is a great place to start. However, God also wants us to do more than just talk with him. There is more we can do and in the next chapter we will go over more ideas.

## MEET TOMMI

D13: Saber Romeo One Four, Davey thirteen airborne.

SR14: Roger 13. Have call from Stepfather. They have located target at Papa Oscar 5-4 and need CAS. Contact Stepfather Six onsite.

D13: Roger. Will contact Stepfather 6 in 10.

SR14: Wilco. Saber Romeo One Four clear.

They say there are no atheists in foxholes. That is certainly true of my team. I have the greatest respect for the other five guys in my unit. Every one of them is a devoted soldier and Christian. Course we aren't actually in a foxhole, it's more of a dry river bed 1.5 clicks from a warehouse we're about to attack. With some local intel, we have located the enemy in a warehouse and have called in an air-strike. It will be 10–15 minutes before the pilot is in range. He'll contact us and begin to drop bombs that will cover our unit's advance on the warehouse.

I'm Captain Tommi and whenever me and the guys have a few minutes like this, we read some Scripture, discuss it and share what's happening back home. We always start with a prayer together. Today Samuael is leading the study. Most of us use our smart phones to access youversion.com/live where Samuael has the Bible verse already setup. Today's verse is James 4:7–8.

*Submit yourselves therefore to God, and he will draw near to you. Cleanse your hands, you sinners, and purify your hearts, you double-minded.*

I love using youversion.com/live[5] because I can also make notes regarding thoughts I have while we're talking about the verse. All the guys can share their prayer requests online and we can refer back to our scriptures, notes and prayer requests whenever we want.

Our whole unit uses the Heckler & Koch H&K-UMP with quick-connect sound suppressor attached as our primary weapon. Every man usually also carries a pistol, some sort of fighting knife and just before we move in, we use flash-bangs (M84 stun grenades). We stay in radio communication throughout the assault with our headsets. We also have high-tech visors that give us a readout of each other's locations transmitted by the unmanned RQ-4 Global Hawk surveillance plane flying overhead. The strategy is that the enemy is plenty shaken up from the aerial bombing that covered our advance. The very next second when our team enters the building, the flash-bangs cause immediate blindness and deafness so they don't see or hear us. It may be temporary but a few seconds is all our elite unit needs to secure the area.

If all goes as planned, we take prisoners but that always depends how badly the other guys want to fight. As soon as we secure the area, with or without prisoners, helicopters will land, ready to evacuate everyone. Hopefully, it'll only be a couple of hours before we're back on base in the pub enjoying the usual praise from our fellow soldiers.

Our unit is one of the toughest around and the whole unit gets lots of respect for our fighting skills and even our faith. We never push our faith onto anybody but we don't hide it either. It's amazing that some of the other soldiers will often come to one of us privately and ask us to pray for something going on with them. We have been surprised when a guy who acts tough, as if God doesn't mean anything to him, will come and ask for prayer when he gets bad news from back home. Maybe his mom has cancer or his sister's kid fell out of a tree. We've even had an enemy prisoner share a request for prayer or ask us about our faith. Lots of people have questions about the Bible, and we all try our best to be prepared with an answer. That's why we take the time to get together whenever we can to pray and read the Bible together and discuss what a particular scripture passage means to us. We study to be ready and pray that God's Holy Spirit brings it to memory when we need it.

Tommi is probably the oldest of the guys at 26 years old. Samuael is 19, Isaac and Danny are both 18. Taidgh is 21 and Chris is 22. All the guys enjoy their video games and this MMORPG (massively multi-player online role-playing game) gives the guys a masculine avenue for their regular studies and discussions about faith. It also gives them a place to share their faith. They don't all live in the same town but they are able to keep each other on track while having a good time role playing in the games. For some reason, even though the battle is a video game, they all feel like a band of brothers who are always looking out for each other. They really do feel as if they would take a bullet for each other and in the game they have.

## EPHESIANS 6:10-20 ESV

"Finally, be strong in the Lord and in the strength of his might. Put on the whole armor of God, that you may be able to stand against the schemes of the devil. For we do not wrestle against flesh and blood, but against the rulers, against the authorities, against the cosmic powers over this present darkness, against the spiritual forces of evil in the heavenly places. Therefore take up the whole armor of God, that you may be able to withstand in the evil day, and having done all, to stand firm. Stand therefore, having fastened on the belt of truth, and having put on the breastplate of righteousness, and, as shoes for your feet, having put on the readiness given by the gospel of peace. In all circumstances take up the shield of faith, with which you can extinguish all the flaming darts of the evil one; and take the helmet of salvation, and the sword of the Spirit, which is the word of God, praying at all times in the Spirit, with all prayer and supplication. To that end keep alert with all perseverance, making supplication for all the saints, and also for me, that words may be given to me in opening my mouth boldly to proclaim the mystery of the gospel, for which I am an ambassador in chains, that I may declare it boldly, as I ought to speak."

# CHAPTER 6

## PROTECT YOURSELF

After the last chapter, did you pray about your place in online missions? I hope you did. While prayer is an effective part of protecting yourself, there are more steps we should undertake before stepping out into ministry. I have already mentioned that I am not a huge fan of some forms of downtown street witnessing, but if you are doing this kind of work, hopefully you don't go out unprotected and unprepared. I hope you use the buddy system, bringing along a few others to watch out for each other. I would hope that you don't go out to present God's love without studying his words in the Bible.

As I prepare to write this chapter, I feel a little uncomfortable. After all, I'm writing this entire book to Christians. Not only to Christians, but to Christians who probably are going to attempt to stay away from temptation online. So it doesn't seem that I should need to bring up some of the issues I'm about to share, but I believe that I need to do so anyway. I need to overcome my caution because this stuff is important. Stupidly simple, but still important.

I have attended church since I was about two weeks old. As a teen, I was overwhelmed by the sacrifice God made on my behalf and the love he showed to me. I chose to go to a Christian college after high school and to earn a degree in Biblical Studies. Sometime after I graduated, my fiancée and I had a church wedding and we presented ourselves to God before our family as a couple who would begin a Christian family. Since our children were born, my wife and I have taken them to church regularly.

I don't say this to impress anyone. I too have gone for long periods without doing some of the things I will suggest to you here. It is because I haven't done these well that I know they need to be done. The next few ideas can really help, and if you are not doing them I hope you consider making some changes.

### *Read the Scriptures*

If you are a Christian, you should be reading the Bible--all of it. I first read the entire Bible from beginning to end when I was about 16. I did this as a requirement for a Bible Badge in the boy's group at our church. I later re-read the entire Bible when I was in college. As I mentioned before, I went to a Christian College and at the time it shocked me how few people--even in the religion programs, such as Biblical Studies, Pastoral, or Missions--had ever read the entire Bible. I know this because I asked lots of people. And yet, even though it shocked me at the time, after reading through the Bible a second time, I didn't repeat this action for about 25 years.

One of the best things my wife and I ever started doing together as a married couple was to decide to join in with the annual Bible reading program at our church a couple of years ago. We were to obtain a copy of the One Year Bible (Tyndale House Publishers) and follow the schedule to read all the way through it within the course of a year. The One Year Bible, now available online at http://www.oneyearbibleonline.com, is laid out with sections from the Old Testament, New Testament, Psalms and Proverbs every day. This very action improved our marriage, including our communication with each other. In the years following we have read other one year programs including some on youversion.com.

I spend a lot of time in conversations with people on Facebook and other areas online. Generally I enjoy discussions related to government, business and personal finance. For some reason, the issue of God or religion often comes up in these discussions. One of the very good reasons for this is that in the United States, much of what we feel about these areas has been strongly influenced by the Bible over the years. Some people claim that the founding fathers of this country intended for this to be a Godly nation. I have heard others declare that this is not true. While many will debate whether or not this claim is true, I do believe that all the founding fathers knew more about the Bible than many of us do today. They read it and I believe they were impacted by it as well. This influence from the Bible was a factor in how our government was set up and how laws were created. The same is true for most of Europe since many of the laws there also were developed as a result of influence from the Bible.

Whether you believe the Bible is God's word or not, it has been a very influential book on the issue of morality for longer than the United States has existed. Because of that influence, all people have to decide what they believe about God and about the Bible, including whether they believe God exists.

The problem is many of us haven't read the Bible for ourselves. As I get into these discussions and the issue of God comes up, it's interesting to see the variety of opinions that are displayed. Many come from extracting a small section of verses. I've mentioned it is important to read the entire Bible because all too often in churches we study only a few sections over and over again. We leave out some of the "boring" sections lest we fall asleep in our studies.

People can justify just about any personal belief from scripture if we only look at a short, isolated selection of verses. I have witnessed this in my online discussions. I have to say that at times I have been guilty of doing the same. However, I try to consider more than just a few scriptures, to take in the meaning of a larger passage. Even developing doctrine from just a few books such as John, Acts and Revelation can be incorrect. It is important to know the story of God from the whole Bible.

It was a wonderful experience to read the Bible with my wife that first year and now we continue to repeat the practice every year. Yes, we could go through the Bible individually, but it's nice for us to experience it together. My wife asks questions that I might take for granted if reading alone. I also bring up points that mystify me.

I know, for example, that the scriptures claim that Jesus Christ (God) is the same yesterday, today and forever. In other words, he doesn't change. And yet I have a hard time reconciling some of God's actions I read about in the Bible, particularly in the Old Testament with sections from the New Testament. At first glance, it seems to me that God does change, even that he changes his mind. This blows me away because as an all-knowing, all-powerful being it wouldn't seem that God would change his mind. Certainly not because of anything that a weak, incompetent human would do or could present to God for consideration. And yet there it is in the Bible and if you research it further, scholars can often help us understand why it appears to be this way.[6]

Be prepared! You will be shocked by what you find in the Bible. I have found that I need to read and re-read portions over and over again before some of the small stories that might even be only one or two verses finally hit me between the eyes. No romance, spy, or war novels contain any shock value that has not been in the Bible for a long, long time.

### *Study the Scriptures*

Beyond reading the Bible, we need to study it as well, to read what others have said about the scriptures. Today there are many sermons online, many of which take a book of the Bible and analyze it. Listening to someone who has looked closely at the scripture is a great way to study the Bible. After all, not all of us are going to sit down and read the intellectual books or learn to read Greek, Hebrew, Aramaic and Latin in order to translate the scriptures for ourselves. But we can study along with someone who does.

There are also many other ways to enhance your study of the Bible. One is to obtain a study Bible. Scholars who put study Bibles together do the hard work for you.

College degrees in Biblical Studies are also available. Bachelors, Masters or Ph.D.'s exist in many fields related to the Bible. Become one of those Bible Scholars by learning the history, archaeology and literary backgrounds of the scripture. I enjoyed the variety of classes I had while getting my Bachelors in Biblical Studies.

### *Reflect on the Scriptures and Journal*

Reflecting on the words you have read and studied is also a great practice. This is an area I have never been good in but I know that reflecting and keeping a journal about your thoughts can be very helpful. It's important to read the Bible but also to store it in your heart and consider what it really means and consider who God really is. As Christians we need to contemplate what God's desire is because without reading, studying and reflecting on the Bible, we tend to depend on ourselves.

In our online ministry, it is extremely important that we don't rely on our own desires. After all that's what is getting so many into trouble online. On the Internet, we are anonymous and if we believe actions online don't reflect reality, we can get into dangerous territory. However, God doesn't see us anonymously when we are online. He always sees us exactly as we are. As we encounter people online we will need God's words close to our heart so that he can remind us who we are and give us the correct words for the current discussion. We need to study the scriptures to keep our hearts pure and to effectively share God's love with others--both in the real world as well as online.

### *Real World Meet-ups and Small Groups*

In many churches today, we see a trend to form small groups (something that actually is the oldest practice in the history of the Christian church). Sunday morning services can be a wonderful encouragement, but we need to make things real for us throughout the week.

I have often wondered why God says he wants people to fellowship together and why the Bible says that if two or three people are together, he is in their presence. Doesn't God realize how messed up church people are? If I were God (and let's be thankful I'm not), as soon as a person became a Christian I would whisk them away to a mountaintop somewhere and train them on everything they needed to know to be in the club. I would give them the decoder ring that would help them understand what the scriptures mean and, just for fun, it would also have a male/female setting so that the person could better understand their spouse. I would hope that my wife, for instance, had a ring to help her decode every signal and word I spoke as well. Wouldn't that be awesome? Seems to me it would be, but God didn't do it that way. There is no decoder ring--at least I didn't get one, did you?

Instead, God wants us to not only pray and study his words to us but he also wants us to hang out with people, even those who frankly may not be any better off than I am. People have problems. Being around some of them makes me have to pray often. They ask for prayer when they get hurt, when they have a relationship issue, when their business is doing poorly or maybe when their kid has a problem in school.

But it's not just about praying for them. With these people in my life, I have them to pray for me as well. I have to pray that I don't mess up like I did last time when they shared with me what was going on in their lives. I have to pray about my responses and my judgmental attitude and that my heart doesn't become too bruised or calloused by listening to all the junk that goes on in their life. And if praying isn't enough, I have to study the scriptures just to have the right word to pull out when it is needed or to reflect on when I'm listening to them. Does this sound familiar?

Being around needy people is a lot of work. I have to learn how to relate to them on their terms without offending them or maybe even getting shot (okay hopefully not shot). I have to adjust my behavior for these people. I have to adapt to the way they do things. I have to put up with their viewpoints and opinions about almost every issue under the sun. Can you believe God would do this to me? To us? Oh wait. Maybe that's part of the point. Maybe he has me around these people so that I will pray, I will study the Bible, and I will adapt myself and change my behavior because of the mistakes I make. I have avoided this issue for a long time because it's a difficult one.

And yet, I have been in some really great small groups. I want to thank everyone who has put up with me in a small group. I have also spent a great deal of time not being part of a small group. On my own, I become too worried about my personal issues, my life, my kids. While it is important to keep these things in mind, I also need to be involved in other peoples' lives. This helps me keep things in perspective. It's good for me to care for others.

This is true for any gathering of people--not just in church--which is probably why people worldwide are breaking into smaller groups all the time. We form groups for hiking, photography and groups related to our children. We can benefit greatly from being involved in special interest groups. But unfortunately, sometimes we become too busy with groups unrelated to our spiritual health and our relationship with God.

Yes, we are personally responsible for our relationship with God. I also believe that God does not intend for our relationship with him to be a solo act. I don't want to add more stress to your life and cause you to say, "Oh great. There's yet another thing I have to do". Rather, I want to encourage you to realign your life so that you are doing what is really important. You may need to drop a few things so that you have time for prayer, Bible study and participation in a small group of like-minded Christian believers. You may even find that in some small groups, you can accomplish all these things at one time. Talk about multi-tasking. It is nice to have somebody out there besides your immediate family who care about you.  Particularly important to be in a small group if your family doesn't care about you. It's nice to have somebody listen to you and look out for your well-being. This is important and realizing that it is has made a big difference in my life.

As we look at beginning a ministry online, I hope you realize why small groups are important. You may get involved with people online who pull you away from God's design and purpose. If we don't have a Godly support system, any of us can be drawn into things we didn't intend to do. As we talked about in the chapter on The Real Me, the people you come across are people who often did not set out to do unethical or sinful things. The apostle Paul himself didn't want to do wrong things, but when we get busy with life, it's all too possible to let the important things go and head in a direction we don't intend to go. As we all begin to explore our lives online, it's possible to neglect the important things there as well. Even though it doesn't seem real online, there are places and circumstances in that realm that can injure our souls. We need the support of others who believe as we do and who will be praying for us as we approach a particularly difficult person in Second Life or on Facebook, or any other places we may find hurting people to pray with and who we can help.

### *Online Meet-ups and Small Groups*

I've been talking about small groups in the real world, but what about virtual small groups? As more people begin to realize they have a Christian witness online, it is important they not attempt to go it alone. Would you walk into a bar known for prostitution by yourself to begin witnessing? Maybe you would but it's not recommended. Just like we need small groups in the real world, we need to form small groups online. Like the "band of brothers" I mentioned above that meets online, it can be very beneficial to study scriptures and pray together online. These young men just came up with a fun way to meet. If we walk into a virtual bar in Second Life we may need others with us. If we get involved in groups or start discussions on Facebook, it's nice to be able to call in somebody else to participate or at least be reading along on discussions we have with others online, so that they can advise us. I get that kind of support all the time while I'm online.

There are many places we can meet for study or prayer online and I hope we see more of these groups forming. Hopefully, your real-world church has a variety of small groups to get involved with. I hope the same for the church online. I hope that we develop many different virtual groups so we can begin to coordinate our efforts and become a force online that shows God's love to everyone we meet. I hope that iMChurch, which I mentioned before, will be a place where, among other things, you can find online small groups that fit your time and interests, many of those will be in the local church if they chose to be. My hope is that someday we can also provide materials for Bible study and small group discussions. I would enjoy being a part of making this happen. Like Moses in the Old Testament I was frightened about becoming a large part of making this happen but since God is calling me in this direction, I knew he would pursue me until I became involved.

My wish for you is that you will find small groups and meaningful relationships in the real world as well as online to fortify your efforts in both places and to keep you moving toward a closer relationship with God.

But if anybody does get a decoder ring, let me know. I'll have to talk with God about that one if I find out you did get one, and I didn't.

### *Finances*

You might ask what finances have to do with online ministry. Please don't tune me out like you might when your pastor talks in church about the Biblical perspective on money. If you are a Christian and you say you believe in the Bible as the word of God, I have a shocking truth for you--the Bible talks about money quite a bit.

I'll start by saying that once again I have failed. In this area, my record isn't very good. At least I can no longer be self-righteous about it.

Why do you think people don't want to hear about money in church? If I could get Warren Buffet to come to church and talk about money, the place would be packed and satellite areas with screens would have to be set up so everyone attending could hear him talk. I think the same would probably be true if I could get the CEO of Chase Bank to come and talk about money. Many would want to hear what either of these people have to say about money, wouldn't they? Why? Is it because they care specifically about the financial well-being of our families?

Do you think Chase Bank would refuse someone a loan to buy a new car if they thought it would be difficult on the family's budget? Would any of Warren Buffett's Berkshire Hathaway companies refuse someone trying to buy insurance or a box of chocolates when they truly needed the money for groceries? I don't think they personally care about your budget, only about their companies' bottom lines. That doesn't make them horrible in my book though.

My point is that if a pastor were to announce ahead of time that he was going to talk about God's perspective on money, many would say, "Oh no, a tithing sermon" and might not make it to church that week (just like I would try to avoid Missions Sunday), in spite of the fact that the opinion of the one who created the universe is more important than Warren Buffet's.

I like the Biblical perspective on money because it's one of the only sources I know that comes from someone who actually cares about my own family's financial well-being, who cares whether I have joy and blessings in my life, particularly related to my management of finances. You see, the Biblical perspective on money comprises much more than paying 10% of our incomes back to God (his money in the first place). It has to do with investing our money and tries to protect us from taking out loans because the Bible says that if we do, we will become slaves to the lender. That is true.

If we actually open ourselves up to the Biblical perspective on finances, we would see that there are all kinds of good things in the Bible about financial happiness. We would also understand that we have some responsibility in not only giving back at least 10% of our hard-earned cash, but we have a responsibility to live on far less than 90% of our income so that we can do a little good in this world. The Bible says if we take care of the poor, we will always have plenty. But it's not all about others, it's also about you. This excites me and frustrates me all at the same time. As I said earlier, I have blown it big time on this issue. At one time, I had an extreme admiration for money and the people who make lots of it. In the 80's, I started buying Forbes magazine. I enjoyed the business articles because they talked about things I was learning in some of the business classes I took. I also enjoyed every October when a list came out that was called "The Forbes 400", a list of the 400 wealthiest Americans in the previous year. For me, it was like baseball cards are for other people. I admired those people for what they had accomplished, even if I didn't always admire everything about the people. I wanted to start my own business and I wanted to be considered successful.

Eventually, I did and made a decent income for a number of years. It felt really good and I felt pretty smart. So if you ask me why I am interested in what the Bible says about money, I would say that I found out I wasn't so smart. I kept making more money and then I kept spending more. Turns out I didn't understand a thing about money. Maybe there are people who are not Christians who do understand money but I'll tell you what I now realize is that even if those people don't profess Christianity or a belief in God, many of them are mostly following Biblical principles of money, even if they don't realize it. I believe the Biblical principle of giving works even if you are not a Christian and even if you don't give to a church. Of course, they may or may not be following the principle of choosing God over money.

So here I was, successful in my computer consulting business yet barely staying ahead of the debt collectors. Then the market hiccupped and the delusion I lived under was gone. I was giving quite a bit of money to the church but it wasn't 10%. Also, I wasn't following other

Biblical principles I should have been following. My loss happened quickly and I had to sell the great house I had bought only a few years before or else go into foreclosure. I found out how close to the surface my problem was when I quickly lost everything I had worked for over the previous two decades. Unfortunately not long after, many other people went through the same problem I did. Even as I write this, the United States and most of the rest of the world is in a financial crisis. I don't know exactly how to help things on a macro level but I can tell you what has made a big difference in my household--understanding the Biblical perspective on money.

Given those conditions, why would I start paying 10% of my remaining income to God's church? I found I didn't have any other solutions and God promised to bless me if I did. I didn't know exactly what kind of blessings to expect, whether it would be money or something else. I no longer believed I had all the answers. All I had was a promise. I knew my way didn't work so I looked into God's way. Even though I made far less money than I once had and felt ridiculous for sending in my small checks to the church, I did it anyway--and made sure it was at least 10%.

The biggest reason I recommend you do the same is because I believe you will witness a miracle. Over the previous several years, I hadn't had any money left over each month--it didn't matter if I made $10,000 or $2,000 a month. I still didn't have any money left over. So it seemed to make no sense to give away 10% and then expect to still get by. But by being obedient, I witnessed miracles. I use the word miracle more loosely than some would I suppose but I see little things happen and I don't know, how come they make life better or what to call them otherwise.

One of those miracles that I do understand was attending Financial Peace University by Dave Ramsey. Today I still can't explain the miracles but after being faithful and after getting some great teaching, I started to have money left over. I am working on the fully vested emergency fund that Dave talks about and I'm blown away at how much is in that account now. Back when I made good money, if I had been following God's plan, I probably could have had 10, 20 or 30 times as much money socked away in an emergency fund. It took years of concern and doubt on my part, but I kept to the plan. It is a miracle what God is doing in my life right now and, no, it isn't all about money. But the miracles related to this area of obedience are now astounding to me. I feel God saying, "It was no big deal. You just had to obey."

So what does money have to do with online ministry? Maybe if I can realize financial freedom God's way, not thinking it came from a superior intellect on my part, then this will free me up to focus on more of what God's plan is for my life. For me, that meant this book (which has been brewing in my mind for a long time) finally came together and it became clear and I actually had some words to put on paper.

For you it could be something different. Hopefully, following the financial principals found in the Bible will put you into a position to minister to people online without being stressed about paying your bills or whether you can afford to buy groceries. Hopefully, you'll be able to focus more on the people you meet in the real world and online and will be able to focus on answering their questions. Hopefully, by not being distracted with worry, it will be easier to recognize potential harm that may come your way--or I should say that will come your way.[7] This is just one more piece of the package you can use to protect yourself online and, for that matter, in the real world. The Biblical perspective of money has everything to do with being a blessing to others, if you are, you will be blessed as well.

If you are praying, reading and studying your Bible as well as meeting with one or more groups of Christians online or in the real world, as well as living according to God's financial plan for you, I think you will be okay. The only thing you now have to figure out is how you are supposed to put all this together. I have some answers for you in the next chapter.

# MEET CHARLES

"Yes, Charles, I personally know Nathan. He and I dated a long time ago. We went to the same church. Maybe you have heard of the now defunct On Fire Ministries..."

Angela's note went on but Charles had stopped reading. Nathan had been in a Christian cult? He leaned back in his black leather desk chair and ran his hand across the end-of-the-day stubble on his chin, staring at the blue glow of the computer monitor on the ceiling of the otherwise dark room.

Before heading for bed, he had decided to check Facebook one last time and found a short message waiting for him from Angela, a woman he met in a discussion on Nathan's wall. He had never met Nathan in person either. He was a friend of a friend whom he discussed mostly politics with online. One thing he had noticed in their discussions--or sometimes arguments as the case may be--was how negative Nathan was about Christianity. Every time the subject of Christianity or any other religion came up, Nathan was quick to cut down the person who had mentioned it. He claimed to believe in a God who created this world but also said, "It's abundantly clear that this God has abandoned all of us and leaves us to our own devices" and "The Bible was written by a bunch of guys long ago who were set on the task by a ruler who thought that religion was the best way to control the population he now ruled over."

Very interesting. Maybe there's a connection, Charles thought so he did a Google search for "On Fire Ministries" to find out a little more.

He found the website of a former member and located a backgrounder page regarding On Fire Ministries (OFM). As it turns out, one of the reasons that mainstream Christianity considered them a cult was because they denied the traditional teaching of the Holy

Trinity in favor of a doctrine centered on the "Oneness" of God. He is still trying to understand the doctrinal differences because even the belief in the Holy Trinity believes that God is One God. But of course Charles believes there are also three parts to this one God. The doctrine of Oneness does not! He has also found out by asking questions that Oneness people don't necessarily like the term Oneness. To confusing!

Charles continued his research and found a Wikipedia article that described one of the practices of this group.

"During church services, members were instructed to find a partner of the opposite sex but not their spouse, known as a 'flame.' By staring into each other's eyes and touching the temples of each other, a process known as 'lighting the flame,' partners were told they would actually experience God through each other's touch, and they were encouraged to look with love into their flame's eyes in order to express their love to God. Throughout the week, both in and out of church, members were encouraged to spend time with their 'flames' in a kind of quasi-dating relationship. Physical intimacy often accompanied these times together as flame love was taught to be more intense and more desirable than marital love. A former member has stated that the connecting experience was so intense that she and other women would experience orgasms without having true sexual contact with their flames. But at the highest levels, sexual relations were not discouraged. This was seen as a completion of God's instruction that we have love for one another."

By searching in various other places online, Charles discovered that OFM had grown quite big by the mid-80's, reaching thousands, and the church services were ahead of their time in the atypical contemporary music they played rather than hymns. The music was loud and the band used electric guitars, drums and electronic keyboards instead of traditional piano and organ music. They also had spiritual dancing involving partners.

During the Google Search, Charles found a couple of blogs written by former members of OFM reflecting on their experiences. One of those bloggers had left the church when he began to notice strange teachings being put into practice. He already had a background of Bible-based teaching before signing up for classes at OFM's Bible Teaching Institute, so when these new practices were introduced, he had a hard time accepting the teachings. He wanted to trust his new pastor but something didn't sound right and he, along with a handful of other people, started to ask questions. He said that some of these people were kicked out, while others left because they didn't receive satisfactory answers. They were just told to accept where the pastor was going, that God had put him into this position and that scripture says we must follow the leaders he has appointed.

The blogger talked about how after things began to fall apart, some who didn't have a more solid base in the Bible didn't know where else to turn. He knew of many who walked away from the experience with a distrust in God. Their view of scripture was clearly shaken and now they had to figure out how to rebuild their lives. People like Nathan, Charles thought. Some of the church members had previously been hippies involved in the drug culture and the teachings of "free love."

As Charles thought about what he had found out, he realized that one of the sad things that had happened was that in spite of the fallacies inherent in the organization, OFM had found a way to reach the drugged-out hippies and introduce them to a relationship with God. The people also received a lot of teaching from the scriptures. Even though some of the teachings were twisted by incorrect doctrine, there was still a fair amount of good teaching being given by very knowledgeable professors.

In the end, the errors in Biblical teaching finally caught up with the group and the church imploded from the inside out. From other reports Charles found on Facebook, the imploding started with the senior pastor, a deluded but charismatic man whom the church members followed as far as they possibly could.

Charles reflected how people like Nathan had done the best they could when their connection to the church, and thus to God and his scripture, fell apart. Nathan had found his own way to sort out this teaching with his continued love of all music and crippled faith in God. Charles began to understand why most of their discussions revolved around Nathan's mistrust of business, church and mankind in general, no doubt a result of mixing his counter-culture beliefs with his bad religious experience. It was apparent that Nathan no longer knew who to rely on and had transferred most of his trust to believing that the government should take care of people. After all, he probably thought, other institutions such as churches and businesses were not capable of it, or they clearly just weren't interested. In fact, Nathan had once argued, both churches and businesses seemed to be more interested in the financial bottom line than feeding and clothing people or providing for their basic health needs, a fair point.

After finding all this out, searching the Internet for information and asking some questions from others on Facebook who had an experience with OFM, Charles could not help but form a different opinion of Nathan. How could he be so critical of the man? After all, what would he have done if his church had slowly started to teach non-Biblical doctrine? Would he have noticed right away or would the first step have made sense, and then the next step also made sense? Even though Charles had a religion degree from a Bible school, would he really have paid close enough attention or would he have gotten caught up in the worship? If the women in the group were all ugly it may have been easy, but what if they were cute? He shook his head. Then it might not have been so bad to dance with them and be seen around town during the week, all the while being obedient to the teachings of the church and the pastor who God had put in place.

After all, the Bible teaches us to respect authority and to listen to leaders placed by God.

Charles' change in attitude toward Nathan didn't mean he would agree with everything the man had to say; he would still challenge the views Nathan held in relation to politics and religion. It just meant that Charles would respect the man behind the statements and do his best to not belittle him. Nathan might have different views and Charles may not have ever met Nathan in person, but Charles had met people in real life who had similar political views to Nathan's. These people hadn't been in a cult. In fact, they had gone to church with Charles. They were good people and Charles loved them, so how could he do anything less for Nathan if he ever hoped to be of any influence? Charles didn't hope to change Nathan's political beliefs but he did hope that one day Nathan would see the reflection of God's love in his words and in his nature. Charles did hope that someday he could help Nathan to see a value in serving a Biblical Christ again.

(Continued in Chapter 9.)

## EPHESIANS 2:10 ESV

"For we are his workmanship, created in Christ Jesus for good works, which God prepared beforehand, that we should walk in them."

# CHAPTER 7

## ARE YOU IN SHAPE?

I don't want you to feel guilty like I did after my missions class by suggesting that if you are a Christian then you must have an online ministry reaching the lost utilizing this technology provided to us by God. I just want to say that if you're online and you are a Christian, act like it. Seek out Christians online and in the real world to help you be a better Christian in both places.

But if you have read this far and you feel that God is telling you to do something in this area and you want to begin an online ministry, you might ask "What should I do?". I challenge you to examine your online behavior, your interests, and your skills. I have shared with you ideas of potential ministry through the stories I have told about the online world. Maybe one of those areas grabbed your heart and God made it clear that you could do something about this. I believe that God will put the idea of being a Christian witness while online on the hearts of more people than just me. I hope so because if it's up to me alone, then the lives of people on the Internet are not going to change much. No one person has enough interests, skills or time to be everywhere on the Internet and I don't believe it needs to be that way. I think that if everyone who calls themselves a Christian acted like one online then we would see more people's lives change for the better.

### *What is Your Shape?*

I believe God made us all different and he did it for a reason. He shaped us for a particular purpose. I don't use the word "shaped" randomly. You may have taken a test or series of tests designed to determine your spiritual gifts. This is one way you could focus in on what your gifts are. What are your skills? What are your talents? You have a special purpose in life and can apply that in many ways. Hopefully you're able to explore your skills in your daily work or profession. You should also be using these skills in your family and absolutely be using these skills in your local church. These are the same skills you can be using online for God.

To some of you, this may sound like common sense information. But it could be you're like my wife and me and after many years of life and raising two wonderful children, you look at your life and it didn't turn out the way you were planning or even hoped--or possibly some things turned out better than you imagined. But having been on this planet a few years now and watching others, I have seen some who are not spending their time where they are strongest. Some people I have met don't even seem to know what their strengths or talents are. Maybe you are a person who believes that one should focus most of their time on their weaknesses, trying to improve those areas and believing this would make one stronger. The problem is that when we spend most of our time trying to improve on our weakest areas, we will most likely spend years spinning around and around getting nowhere. It's like our life is a tire in a muddy rut with the wheel spinning around but the car is standing still. I don't believe that God gave us talents and specific skills so that we would NOT use them but spend most of our time on areas where we are weak? I believe that God gave mankind talents because he wanted us to develop them and make them even better.

One example is Michael Jordan, the famous former basketball player. At his prime, he had incredible skills in playing basketball. He could jump, he could shoot, he was just awesome. He was well known because he identified those talents and worked on them repetitively. He shot balls at the basket over and over again for years to become even better. Let's imagine that years before a basketball star with talent equivocal to Michael Jordan's took an art class in high school and found out he was lousy at drawing. He thinks, you know I'm weak at drawing. Maybe I shouldn't shoot so many baskets so I can have a few hours a day to develop my art.

It's possible that if this talented basketball player spent hours a day for many years practicing and learning to draw, that he would become a halfway decent artist. But if he practiced art for hours a day and only practiced basketball once a week, I'm guessing we would never have heard of him for either skill.

How about you? Are you working in your areas of greatest skill and loving it? This concept hit my wife right between the eyes a few years ago and she decided to do something about it. For one thing she decided to make changes in her own life so that she was doing what God had planned for her.

***Is Your Heart in It?***

I like to come up with new ideas. A few years ago I had an idea that I believed could expand my business. The problem was, due to the way I was running my business I didn't have the time to make the change. The beauty of this idea was that to develop it one didn't have to be a technical person. So I thought I could continue the core business and my wife could develop the new idea. We branded it, developed the new services and had a website designed to show these services.

But because of my workload, my wife had to start marketing it. She made phone calls and changed the website to reflect the results of her market research. She really tried to dig in and find a simple way to explain these new services and their benefits to average people. She got on the phone and started calling companies. She worked on this part-time for a couple of years and at the end of that, we had nothing to show for her efforts. The problem was that her heart wasn't in it. This wasn't her talent. She didn't want to disappoint me but working on this new aspect of the business drained her. It did not energize her. She worked up the nerve to talk to me about it and because I love her more than my ideas, I released her from that obligation.

Something I should have realized earlier is that part of what has always made my wife tick is the potential in children. Ever since I have known her, she's been involved in children's ministries--teaching Sunday School, being on the church staff for children's ministries, running a summer camp program (Vacation Bible School or VBS). So she decided that it would be awesome if she could help young kids realize that God has made them with specific talents that they can use. She said, "I thought, wow! What passionate lives people could lead if they applied this principle with Christ at the center at a young age instead of waiting 47 years? We could change the world!"

She wanted to help develop the kids and teach them to spend most of their time in the areas of their giftedness, To help them be the Michael Jordan in their field. So since then she has started a nonprofit organization called ETC4Kids.org.[8] The E-T-C stand for Experience Talent Camp. Her dream is to help all kids, even those that can't typically afford to attend camp. She wants them to experience different areas of talent when they are young, discover their talents and learn to develop the talents that they have and, thus, minimize time spent on their weaknesses. Wouldn't that be awesome? She wants to help local churches make these camps a reality, camps that will take many different shapes and forms and may even go beyond traditional summer camp ideas. I think she is awesome and I look forward to seeing what she will do with this concept. The camps are part of a larger idea she has for "kids" of all ages, so that we all understand we are God's masterpieces or poiema in the Greek language of the Bible.[9] That too will be interesting to see what God has in store for her.

### *Do What You Love*

I don't know about you but even though I may have known that God made me special I haven't always been able to get away from focusing on my weaknesses. We should both do better, you and I. Maybe my wife can help us. I will let her write her own book on that though. For now what I'm trying to say is that God may have given you skills, or passions that you can explore online and while doing so, be a Christian witness to others. You may be like me and enjoy political and business discussions. On the other hand that may stress you out completely as I have found out it does to some. You may be an excellent photographer, or you may be an artist, or you may be one of the best players on World of Warcraft. Whatever it is, you can do it online and make a difference. You may have a passion for a topic which you can write and contribute on Wikipedia or some other website. Online there are many different communities and the Internet is changing the concept of what it means to be local. We can interact on a daily basis with people who live across the country, around the World or maybe one block over. All of these people are as close as your nearest computer.

Don't do something you think you have to do, do something you love doing. It's your time online, you can do what you want. If we would all do that, and if we would act like Christians while doing it and share God's love (rather than his wrath), we could be an online force for change in others lives.

***One Big Church***

One of the biggest problems we have as the Church, I believe, is that we don't act like one big church. We act like a collection of anonymous individuals--both online and in the real world. If we did work together, some of us could be the eyes to see what is going on, some could be the ears to listen. We would have brains to figure out what to do, as well as hands, arms and feet to make things happen. Online, I guess it would be the fingers that would make things happen. If we were to act like a global Church every day of the week then we could literally change the world. Not only would we share God's love, but we might also solve problems in the world such as hunger, health care, and relationships. You name it, in the whole body of Christ we have people skilled in everything you can imagine.

Some are blessed with financial resources. That too is wonderful and there will be places for them, but there are many solutions that don't require money. Sometimes we do throw money at a problem and wonder why it never is solved. The reason: money often isn't the solution. This is a hard thing for a U.S. citizen to say but it's true. In the United States we tend to throw money at every need and unfortunately, when we do, we sometimes make the problem worse. Sometimes the better solution is just to listen to hurting people. Other times encouragement is needed. And of course, all around the world people seek companionship. Sometimes a solution requires muscles. We may need somebody who can carry furniture to help a friend or acquaintance move.

Some problems require a certain level of experience. I may have the experience you need or the knowledge required to solve your dilemma. You may be having a problem with your children that I already went through. Maybe my experience will be slightly different and my solution isn't your solution but I can still listen and let you know that many people have made it through the same problem you have and they survived. There are many support communities that operate this way in life, and online we can do the same. If you have a web-cam, it is possible to see each other's faces while talking, just like John's prayer group did in beginning of Chapter 5. At other times such as when we chat online without webcams, it's easier to share the real problem than to make an excuse or put up some sort of wall for protection like we do when talking with people in person. Being anonymous can be a benefit here.

The bottom line is that all of the tasks you can think of may not be easy for you to accomplish, but they may be for someone else. Perhaps God made us all different so that we can solve all the problems in the world--but we have to work together to do it.

Of course, there are some problems that none of us can solve. These are core spiritual issues. However, we Christians do know a man who walked this earth and who knows the right words to say. We know Jesus and we can tell people that our relationship with him has made all the difference in our lives. If we are open and honest with people when online, they may even listen to what we're saying, even if they're not ready to accept it for themselves. In my opinion, that's fine too. After all, when it comes to washing sins away and forgiving every error, this is neither my strength nor yours. Only God can handle this one. We may or may not be talking to someone online when God does his work in their lives, but we can still be a part of the process.

Do you believe that a person can become saved through an online witness? Do you believe a marriage can be saved rather than damaged by surfing the Internet? Do you think a teenager can be talked out of having sex by seeking answers in a chat room? How about using the Internet to talk a young person out of committing suicide? Do you believe lives and bodies will be healed because of the efforts of Christians online? I do and I think it's time we tried. We can setup websites where people in need can come for help. We can setup information sites for people who want to know the messages of the Bible. Those could be big efforts but I believe we can make it very simple and I will get into that more in Chapter 9.

There are some like "To Write Love On Her Arms"[10] an organization which is already trying to help people online who have low self image and contemplating suicide or other personal damage. Don't forget there are people online who are trying to cause people to harm themselves in many ways. They are already there and we need to combat that force. It's going to take many of us and maybe even ALL of us to be Christians online. We may even find someone to help meet our needs while we are there because, after all, we could all use a little help from time to time.

What is your place in all of this? If you have been doing what I suggested earlier, praying, then you may already have an answer to what you can do online. If not, keep praying. I believe God will eventually give you an answer.

## Meet Travis

It's cold outside, gas fireplace turned on... check, cup of coffee... check, comfortable chair... check, reading material... check. Travis is ready to read the articles his friend Sandy Dummont wrote about Wikipedia. She wrote both articles for a small Christian publication "Occupy the Church Gazette". Sandy interviewed Travis for the articles and he enjoyed sharing this passion of his for contributing to and editing Wikipedia articles. When she first asked Travis about the interview it was for only one article. Sandy wanted to let the readers know how Wikipedia works, how information is collaborated and edited. As they talked, a second "darker side of

Wikipedia" article emerged. The fireplace is finally warming up the room, and the coffee is waking him up, as he begins to read the original article first and save the "darker" article for the second cup of coffee. Of course he was reading and paying particular attention to the sections where Sandy quoted him.

Travis says, "Wikipedia is sort of like a social network such as Facebook but for people who want to participate in creating a massive online Encyclopedia". "There are no restrictions to get started, anybody can contribute to a Wikipedia article. Contributions can often be completely anonymous, unless a page has restrictions. But, if you really want to get involved the best way is to sign up for an account." "When you set up your new account you can fill out information about yourself on a 'personal profile page'." Some people setup their user account with an anonymous name but Travis uses his real name for his profile and on it shares information about his Christian beliefs and his faith in God.

In Wikipedia anybody can start a new page for a topic or add information to an existing topic. Travis explains, "Everybody who contributes to a particular topic on the page decides in a collaborative way what stays in the article and what should go, change, or be moved to another location." "Some topics have a lot of detail and become fairly long, so when one of the contributors gets a chance they will break it down into smaller sections to make it easier to find information on the page. Sometimes even a section becomes too long and the decision is made to give that section its own page. A new page is then created and a link in the original article will direct the reader to that new page. For example on the Crusades main page we have sections for each major campaign. On that page you will see a brief description but you will also see a link to another page which addresses that campaign in detail."

So Travis how does everyone keep track of what has gone on regarding a topic they are viewing? "Well, there is more than one page associated with the main article, one additional page is the history of changes made and by whom. And every article also has a talk page. This allows people to discuss what should and shouldn't be contributed to the article. Viewpoints or defenses for material can be expressed here."

Travis is considered an expert and major contributor on Wikipedia particularly in the area of the Crusades. As a Christian he feels it is important to be honest about the good and the bad in regard to this sometimes not so great time in church history.

With so many contributors how do they keep everything organized? And how do they resolve conflicts or arguments? Travis says, "Of course disagreements are part of the process on Wikipedia, everybody can contribute but as people build a reputation of respect they may be asked to become a higher level of editor, someone who is supposed to resolve the conflicts." "A good editor has to allow the article to present alternative viewpoints. Different viewpoints or theories can be given their own section or some may be controversial and moved to a new section identifying them as such."

Good solid article. Travis thought to himself, "Sandy, you should be proud". After a moment to stretch and head to the coffee pot for a refill Travis settled in for the second article:

**"Don't feed the trolls!"**

**by Sandy Dummont**

In last month's article we talked about the details of Wikipedia. You met a friend of mine who edits articles on Christianity for that website and primarily gives his expertise to articles about the Crusades. One thing we didn't talk about is that Wikipedia has Trolls. Very much like the grumpy creatures that hide under bridges and wreak havoc on any who dare pass by, these Trolls cause trouble for the contributors on Wikipedia. Travis told me about one of his more pronounced encounters with a Troll.

"Trolling" is defined by the website as "a deliberate and intentional attempt to disrupt the usability of Wikipedia for its editors, administrators, developers, and other people who work to create content for and help run Wikipedia."

Okay, good start then some repeated background about Travis how he developed a passion for the Crusades in college, and a little more about the editorial process, etc.

Travis related to me his personal experience with a troll. Her user name on Wikipedia was LuciSix66 (whom we'll call "Luci"). I tried to see if Luci would be interested in being interviewed for this article but she has not responded to any requests so far. So we only have Travis' side of the story.

Travis reports, "I first met Luci about 11 months ago. Of course it was on one of the pages for the Crusades, off-hand I don't remember if it was the primary page, but most likely another page. She made an addition to the page that was very emotionally charged against Christianity and had no historical basis that I was aware of."

Travis has mixed feelings about the Crusades himself, "there were things that happened that don't make the church look very good", he says. "That's why I want to be involved, I don't intend to glorify the Crusades but I don't want to over-criticize them either." "At first I wasn't sure what to do with her addition but because of its nature I couldn't leave it in the location she placed it. It was inaccurate. So I deleted it and wrote a note about it on the talk page suggesting we find a more appropriate location for this information."

According to Travis Luci then added it back and continued to do so a couple more times. Travis wanted to proceed carefully. Often one of the next steps is to block a user's ability to post on Wikipedia, but he wanted to make sure she understood. He says, "I thought, maybe Luci hasn't seen my comments on the talk page and thinks we have a battle going. So I sent a direct message to her explaining the situation." "Turned out Luci didn't know about the talk page since she was so new." But that alone didn't clear up the matter. Travis and other contributors continued to have difficulties with contributions from Luci. And now that she knew about the talk page she didn't attempt to present evidence for her claims but spent time pestering everyone with seemingly endless and pointless questions. She could look up the answers easily and sometimes they were even in the article or somewhere else on the talk page." Travis says at times it seemed to be almost personal toward him, "She found my user pages and left me comments there as well, putting me down for my Christian beliefs. She was also clearly finding out about other articles I contributed to and disrupting my efforts on those pages as well." "It was frustrating to say the least."

Yes it was frustrating, and of course something Sandy didn't mention in the article is that now Luci and he sometimes discuss topics on email. Luci surprisingly gave him her personal email address lucisix66@gmail.com[11] and they now have a somewhat civil relationship. It has taken months but Luci has toned down some of the troll activity. Travis had warned her that she could be kicked off if she continued, letting her know they could block her by IP address if necessary. Travis sat back in the chair feeling all warm and toasty from the fireplace and the dark roast, took another sip of coffee and reflected on Luci for a moment.

In her email Luci told him that she used to be a Christian, but as a young teenager she had been molested by an elder in the church. She had a reputation for being a little wild so nobody believed her when she tried to tell people about it. A wild teen was not believed over a long-standing, respected member of the church. Travis didn't know if this story was true or not, but if so, it is clear why she lashes out against religion.

Sandy wrapped up the article nicely.

Travis decided long ago the he would not give up on Luci and hopes that someday she will drop her mask and share even more of her heart to the point where she'll be open enough that he'll be able to help her. Because he believes in the power of prayer, Travis prays continuously for Luci and talks some about her in his church small group discussions. It seems clear that although Luci seems to have rejected God as the one true God, at least she is still struggling with that decision. Travis is looking forward to the day when her struggle is over.

Travis again got up to stretch and thought to himself... Amen, I do hope for that day.

# PSALM 70 ESV

Make haste, O God, to deliver me!
O Lord, make haste to help me!
Let them be put to shame and confusion who seek my life!
Let them be turned back and brought to dishonor who delight in my hurt!
Let them turn back because of their shame who say, "Aha, Aha!"
May all who seek you rejoice and be glad in you!
May those who love your salvation say evermore, "God is great!"
But I am poor and needy; hasten to me, O God! You are my help and my deliverer;
O Lord, do not delay!

# CHAPTER 8

## VIRTUAL BEATING

I hope you are charged up and ready to go. Now you finally realize that you can spend time on the Internet and you are not "wasting time." You can be part of the healing process for the hurting people of the world. You and all the others who join with us are a new force on the Internet, more powerful even than the porn industry's huge presence there. After the last chapter, hopefully you now know where your place is, what your passions are and what talents and skills God has put in you that you can use online. You have prayed for the people on the Internet and you have the support of your small group. Hopefully you even began to take the financial advice from the Bible and apply it in your life and are realizing the power and joy that come from living the way God wants you to live, and that is motivating. I know that each one of you has a unique ministry ahead of you. I just feel I should give you one little warning--these real people you are about to minister to may not like it!

### *You're Making Me Uncomfortable*

Have you noticed that Christians don't always have the best reputation in the world? For some reason, the non-Christian world has certain perceptions about us. They know the rules and they hold us to them. At least they think they know the rules. Some people I have come across actually expect Christians to be perfect. And then when Christians are not perfect, non-Christians believe that proves what they thought all along, that this religion of ours isn't real at all. However, even if you were perfect and God's love flowed through you to all around in the real and virtual worlds, you would probably still be persecuted. The Bible says as much.

For some reason, the realization that we have sinned and that somebody loved us so much that he sacrificed his son for us makes people uncomfortable at first. I don't claim to understand why people are threatened by this. All I'm saying is that I believe it is true. On the Internet, where a person can hide anonymously behind a mask, they still won't feel comfortable if it seems to them that you are reaching into their very soul and exposing them. Their perception might be that they wanted to hide and you didn't let them. If this is the case, realize you are not the one doing this to them. It is God's Holy Spirit working through you.

As we have said before in this book, many people will come to the Internet and feel safe to explore who they really are. Some will be pastors, some will be good church people, some will be youth who are still working out what their faith means to them and what to do with temptations. The Internet makes it possible like never before to actually explore who we are or who we may like to be. After all as I've said, the Internet doesn't seem real so people feel free to go ahead and experiment with different personalities online. I recently read a Facebook post from a psychologist friend stating that "this doesn't seem real," meaning the culture on Facebook. I'm sure that since he said it, some people must think that it must be true! After all, he is a psychologist and thus one who has thought seriously about reality. On the Internet there doesn't seem to be any consequences in the virtual world so we can do whatever we want and its okay. And, of course, there are many people online doing whatever they want to do.

It's basic human nature to want to find other people like us, others who have interests similar to ours. Online we have friends who put up with us, encourage us and help us explore certain issues and interests. We help them do the same. There is a lot of safety and comfort as we sit alone on our computers.

And then YOU come along and share a little light, perhaps something that made someone uncomfortable. You may not mean for it to cause a problem, but it does. Perhaps your sharing makes them have to confront their sins. In that case, you just made them realize that they are not living the life they intend--or should--and that makes people uncomfortable, whether or not they are Christians.

### *Do Not Be Surprised By Persecution (1 Peter 4:12–14)*

Some of the people you come across online are no doubt already being convicted by God and you are going to be there at the right place and the right time to help them begin a new relationship with Jesus or perhaps a renewed relationship. You will then see a metamorphosis before your eyes or rather on your screen.

There will be others you come across, however, who are not yet ready. God is working on them and they are fighting it. They may take it out on you. They might ridicule your faith, calling it a stone-age religion. They might say that there is no way God exists or if he did, he is no longer involved in our lives. They may have many beliefs. Some of them will believe they are living Christian lives according to the Bible but might be racist or maybe they will be judgmental, ready to read the Bible to you as if it is a rule book. After all, there are a lot of rules in the Bible if you focus only on those sections. The people you come in contact with may not see the Bible as a collection of love letters written to make our lives more pleasant and joyous, to make us free.

You will find Christians who believe in God but have elevated some things to be more important than God himself. They may elevate money, friends, family or even the earth itself above God. There is nothing inherently wrong with those things and they are, in fact, important parts of life. People will say things like, "God wouldn't allow my child to suffer." Maybe they will say that God wants all of us to prosper financially and that may be true. Maybe he does. But it's not as important to him as other things are, certainly not as important as the main thing, that is, a relationship with him in which we accept both who he is and his role in our lives.

All of us will come across many stories online. Not everybody puts God first, as King David did. But he didn't get it right all the time. He made some whopper mistakes and they made his life very complicated. David, however, always loved God and understood God's role in his life. Read David's words in the Psalms and you will realize its okay to question God. It's okay to be depressed about what is happening. It's fine to make big stinkin' huge mistakes. It may even be an acceptable emotion to want your enemies wiped off the planet as David prayed for in scripture. But it's also important to view God in the right position and to realize that the big choices are his to make. Sometimes we don't do that. You will come across people who are questioning as David did but they may or may not have the correct viewpoint of God. They may even see you as a messenger of God but to them, that's a bad thing. You are going to take some abuse. Be ready for it. It's a good sign that God is at work.

However, not all the abuse is going to come from your anonymous friends online. Some will come from your friends and family. For some reason, whenever we step out and take a stand, some of the first people to give us grief are those who are closest to us. They often form the first wave of rejection. At least sometimes it feels that way, doesn't it? Why are they so critical? Why do they point out the problems? Why will they accuse you of spending too much time on Facebook?

One possible reason they do it is because they care! It might be that because they are close to you and care so much that they want to make sure you are ready for any difficulties that come your way. Granted, they probably aren't sitting there thinking all of this through and trying to prepare you. The truth is, when you take a stand, you may be making them feel uncomfortable and so they try to diffuse the discomfort by putting it back on your shoulders. This could be another reason why they criticize your efforts.

However, realize that whatever their motivation, God has placed them in your life for a reason. He works through them, even if they aren't Christians. Take the time to listen to what people say to you. Don't reject it but instead consider it. I know this may be difficult. At least it is for me. But if you take a little time to try and understand their motivation, your heart will likely soften toward them and you may see that they are hurting also. You may even gain an opportunity to minister to them in the real world as well as the virtual world. You may even begin to care more for them than you did before.

But now I'm getting a little bit ahead of myself. There is still one more chapter to this book and I think it's very important. You just read some hints about what is to come but if you have made it this far in the book you simply must read the next chapter. When you get online, you must give something to all the people you meet and that's what we will talk about next.

# Meet Richard

Charles walked in the door, kissed his wife hello and handed her a large, white paper sack.

Katherine raised her eyebrows in surprise. "What's this?"

He ran the back of his fingers down her cheek and smiled. "I knew you'd be tired so I brought home Chinese take-out."

It was Katherine's turn to smile and she kissed him on the cheek. "Aren't you sweet? Why don't you wash up and I'll get this on the table."

Over a quiet dinner, Charles and Katherine caught up on each other's day. When they were through eating and had the table cleaned off, he retired to the office and flipped on the computer, ready to check in with his Facebook friends and see what they were up to. The first thing he saw was a private message from Nathan.

Charles: What's up with your friend Richard? That guy is a real jerk. Not really surprised though, he reminds me of most right-wing religious people that I know. ~Nathan

Charles shook his head. Most of their discussions eventually came around to politics and Wall Street finances. Nathan didn't trust big business. Rather, he felt they were taking over this country, slowly eliminating the freedoms on which the United States was founded by buying their voice. Nathan had said before that Charles' support of business as a solution to the world's problems just showed how naive he was. And he definitely considered Charles to be one of those right-wing religious nuts who gets their religion and love of country all rolled up into the Bible, jumbled together into one big confusing mess. But Charles had determined soon after he met Nathan online that he would not be baited into an online yelling match with the man, but would calmly hold his position without spouting off. However, Richard, one of Charles' other friends, was a different story. This guy would jump in all the discussions with both feet, call the President obscene names and tell everybody they were going to hell. He didn't even try to be nice or decent. He just blurted out sound bites and accused Nathan and anybody else who didn't agree with him of smoking dope or taking some other type of drug.

Charles sighed and ran his hand over his face, then heard his wife's voice behind him.

"Something wrong?"

"I got a note from Nathan. Richard ticked him off again." He leaned back and laced his fingers together behind his head. "I just don't understand Richard sometimes. I mean, he had his quirks back in college, but he's always been a funny, creative and artistic guy."

Katherine came up behind him and kneaded the tight muscles in his neck. "Well, I know we had some disagreements with him in the past but for the most part, he seems to be a good guy. At least he believes in the Bible and really seems to care about bringing people to Christ."

Katherine turned and sat on a corner of the desk. "You know, he always seemed to treat people nicely back then. And he was really nice the last time we saw him. Remember he said he was working with the kids in church? It sounded like things were going really great."

"Yeah, that's why it doesn't make sense that he's such a jerk online. The in-your-face attitude he took with Nathan was really shocking."

"Maybe he doesn't consider Facebook conversations to be 'real' somehow." Katherine shrugged. "Maybe he thinks that if he doesn't know the people personally and won't have to face them, that it doesn't matter and he can just unleash his fury on them."

Charles nodded and caught her gaze. "You know, it's as if the real Richard and the online Richard are two different people."

Charles sat silent in thought for a few moments before replying. "Katherine, you know I've really tried to develop online friendships with people I don't know personally."

She nodded, brows raised. "And you know I'm not completely comfortable with that."

Charles took her hands in his. "But I'm careful and I tell you everything. To me, it's like a mission field. I'm just trying to reflect God to these people, show them kindness and maybe at some point, through some of my efforts and through other things God is doing in their lives, they'll open up to God and be more receptive to his call."

Katherine smiled and tipped her head to one side. "So what are you going to do?"

"Well, like I said, I've tried to reach out to strangers but I never thought I'd need to reach out online to people I know personally, much less people that I know already have a good relationship with God." He turned back to the computer and read Nathan's note again. "So I guess all I can do is to add Richard to my list of people to show love to online. God loves them both and so should I."

"And that's why I love you," Katherine murmured and she kissed his forehead. With a gentle squeeze to his shoulder, she left Charles to his thoughts.

After she had gone, he bowed his head and whispered, "God, please be by my side when I type at this keyboard. Please use me to reflect your love for both Nathan and Richard." Opening his eyes, he placed his fingertips on the keyboard and with a deep breath, began to type a reply.

## LUKE 10:27 & MATTHEW 5:14-17 ESV

"And he answered, "You shall love the Lord your God with all your heart and with all your soul and with all your strength and with all your mind, and your neighbor as yourself."

"You are the light of the world. A city set on a hill cannot be hidden. Nor do people light a lamp and put it under a basket, but on a stand, and it gives light to all in the house. In the same way, let your light shine before others, so that they may see your good works and give glory to your Father who is in heaven."

# CHAPTER 9

## LOVE AND RESPECT

I'm glad you've stuck with me to the last chapter. You've probably figured out by now that I'm not going to give you "15 steps it takes to make new converts on the Internet." That's just not my style.

The biggest reason I'm glad you made it to the end of this book is that I believe this is the most important chapter of all. I won't give you fifteen steps, but I am going to give you what I believe to be the first two important steps for living your life online (and in the real world actually). I gave you some other important pre-steps regarding prayer, reading the Bible, getting together with other Christians and about radically changing your finances according to God's principles. But the truth is none of those are any good without these last two important steps.

In Chapter 4, I suggested that the Bible contains the answer to the meaning of life. Did you notice I didn't tell you what I thought that was? I waited till this last chapter on purpose. Too often we look at the Bible as a set of rules designed to take away all our fun. Most of the world sees it that way and to tell you the truth, I think many Christians do as well. We have to fight the impulse to make the Bible into a rule book for life. It's in our nature. I believe the meaning of life is found in Luke 10:27. "Love God first with everything you have, then love your neighbor as yourself."

Many people have already talked about the fact that this doesn't just say "love your neighbor," it also says you have to love them like yourself. I believe this means we have to give them a break. Don't be their judge, jury and executioner. There are, after all, other scriptures that talk about not judging people. Now I've heard excellent expository analysis which says there are conditions when we can judge and then we simply have to be aware that the same criteria will be used on us and if we can hold up to that, okay. This could very well be. I'm not saying I go through life without judging anyone but I want to try to limit the judgments that will come against me by limiting my judgment of others. As much as I would like to always be in God's perfect will and reflect his heart, I am afraid that I don't always get it right. So as far as judges go, I'm not the best choice. I will think I have a situation all figured out and then, when my wife gives me her perspective, I just sit back and calmly reflect on her words. Or do I? The truth is, that's often not the case. Instead, I sometimes try to argue with my wife to justify my position. Like I said, I'm not perfect.

***Option #1 or Option #2?***

When I do reflect, I go back to my comments on why God leaves us around other flawed people while He helps us work out the details of our new faith. Maybe it is through community or fellowship that we actually learn more than we are capable of on our own. Conflict in these relationships is inevitable but in any conflict, we still have choices. We can close our eyes, refuse to take in any more information and give our final judgment from only our point of view. That's option #1 and I'm guessing just about everyone who is reading this book has chosen option #1 at some point in their lives. I certainly have.

But there is at least one other option available. I'll call it option #2. You could stop and try to listen to the input you receive from others around you. Often we don't stop to think that maybe God has put them in our lives for this very purpose. You have your opinion right? It is probably an opinion you have developed over many years of taking in data. The events in your life have helped form that opinion, so you have come to respect your own opinion and hold it dear. Maybe one part of loving others as we love ourselves is to realize that others may have a different opinion than we do for good reason... and respect them for that. Maybe they had different experiences in their life that helped to form those opinions. Often we can benefit from their viewpoints at least as much as from our own viewpoints. This usually isn't an easy thing to do but, nonetheless, just because you respect somebody else's opinion doesn't mean you have to change your own.

In the real world, I have had a difficult time with this concept. As I said before, I enjoy political discussions. In the real world I found that when I had these discussions it was far too easy to have my "buttons" pushed by somebody else—someone who wasn't showing any respect for my volumes of reading and research on the topic at hand. In fact, they sometimes tried to make it seem that they were better informed and more well read than me and that was a very frustrating experience, which made me defensive and more eager to hold to my opinion rather than consider whether theirs was even a valid option.

### *School of the Internet*

Years ago I was excited about the Internet. I was excited for many reasons but one of those was because I saw the early Internet as a challenge to help improve my skills in written communication. At first, the primary medium for this was email, followed by instant messaging and then blogging. I don't know if you have ever noticed this but when we write something, neither the tone in our voices nor the twinkle in our eyes comes through, so the resulting communication can easily be misinterpreted. It was so frustrating to realize that, but since then I've determined to become a more effective communicator with written words in addition to spoken words which I had worked on.

Much of my life I have been in the Technology field but I've almost always been in sales in one form or another. To own a business means you must be able to sell. You must be able to walk into a meeting with a new customer and say the right words, not too many words, and make a positive impact. When the Internet became prevalent, many people started shooting emails around. I took that as a challenge to get better at being able to communicate in written emails. When Facebook came out I got even more excited about the ability to discuss things with people practically in real time. I love Facebook because it is often closer to real time than email and yet not quite as immediate as instant messaging. I now had a new challenge. Now I had a forum to discuss my political and business issues with all kinds of people, people I had known throughout my life as well as new people I had never met in real life. This was very cool.

I tend to not post extremely biased views regarding politics (although sometimes I do as a launching point for discussion) on my own Facebook wall, even though many of my business postings lead to biased political discussions, probably because of who my friends are online. However, for most of my political discussions I try to get involved in what people on my list of Facebook friends post on their walls. I want to participate in their discussions, even though I don't always expect them to engage only in topics I prefer.

When I first started engaging in these online discussions, I am ashamed to say that sometimes I reacted much like I did in the real world--my boiling point would be reached and I would lash back at people who disagreed with me. And yet, this was not what I wanted to do. I wanted to be respected for my viewpoints, so I had a choice to make. Just like the options above, I could chose to close my eyes and blast away or I could take the time to respect others' viewpoints and learn from them. Eventually, I chose option #2.

I have some friends in real life and now virtual friends who don't have the same viewpoints on politics or business that I do. Now that I choose to be more calm, I have become fascinated at how somebody could look at the same data that I do, compare the same charts, read the same news stories, and then come up with a different conclusion. I must love and respect them or they will not be my friends much longer. They will not help me to learn more. This doesn't mean I have to change my opinion, but it does mean that if I love them as I love myself, I will give them the right to hold an opinion different from mine.

Sometimes I still feel the same frustration that I feel in a face-to-face discussion but online I can remember my ultimate goal and take a break for a while. I can step back from the heat of the moment and think a little. The more I practice this, the better I am becoming at it. Also, the more I practice, the quicker I can calm down. And the best news is that as I work this out online, I am also working on a skill that helps me in the real world. Remember I'm never going to pretend I'm perfect, but I am improving.

To me, that's how we practice love and respect online--and in the real world. It's about the value you place on other people. Or think about it as seeing others as God sees them. It's hard to imagine God loving Hitler or Osama bin Laden but I believe he did. You see, when we talk about reaching people on the Internet as a Christian, we have to work on our perspective first. We must follow the principle Christ gave us to love our neighbor as ourselves (as recorded in Luke 10) when we interact with people online. We must not simply form our opinions of them by their online activities, such as viewing pornography or participating in bullying, or looking for a hook-up, or even playing games in which people are killed. We must see a person who is just like us--somebody who has sinned--and we must give them a break just like we do ourselves. Maybe we can even participate or at least accept their rationale for doing as they do--just like we rationalize our own behavior.

Some of my friends online who don't agree with my values for business or my views of politics, also don't share my faith in God. They may believe in God but not believe that he sent his one and only Son to die for our sins. They may put a little too much emphasis on views and doctrines with which I'm not comfortable. But I have found that if I don't treat these people with respect regarding their business and political viewpoints, they are not likely to listen to either my business and political viewpoints or my religious viewpoints.

In my world, religious topics almost always come up and this gives me opportunities to share with others in a variety of ways. Personally, I rarely quote scripture to people who are not Christians and who do not view the Bible is the Word of God. I find I'm better off illustrating my point from another piece of literature because they are more likely to listen and, after all, much of it borrows concepts from the Bible. Even people who do not believe in the Bible as absolute truth and who have new age viewpoints of God or call Jesus a self-actualized human being will allow me to pray for them online. I usually don't actually type the prayer out (although I have on occasion) but if they express a problem, such as a medical condition, loss of job or dog going to the vet or whatever, I tell them that I will pray for them. Most people thank me for that. As soon as I say it or maybe even before, I actually stop typing and say a prayer for them. To me that is a huge opportunity to reach out. It's a very simple step but still big. I have had people who I was pretty sure had no personal faith ask me to pray for them. There are far better prayer warriors than me out there but in these situations I do it because I am the one God put in contact with this friend.

Treat people with respect and you will not have to force yourself and your faith onto them. Let them know your story, the difference that God has made in your life and respect the relationship they may or may not have with God and you will, I'm certain, have opportunities to answer tough questions or pray with people at difficult times. What happens from that point on is more up to

God than you, so take that pressure off yourself. You don't have to hit a home run every time you get on the Internet. You just have to keep showing up for practice. This is the important thing.

Can you imagine Richard (from the story at the beginning of this chapter) being able to do that? If he is rude and hateful, do you think people in the discussion will feel like asking him for prayer? He may have different talents than I do but I certainly couldn't jump from an offensive comment into a meaningful conversation about a God who loves people. If I just unleashed my fury on someone, I would have a difficult time switching gears and having a meaningful prayer for that person without making it a condescending prayer. Even though I can't agree with the actions some take in the name of God, there are things I can do--and should do.

This became very apparent to me one Sunday in church when I saw a video about a man named Andrew Rugasira, who was endorsing a concept he called "Trade not Aid" for the African continent. Born in Uganda, Andrew trained in London as an economist, then he returned to his native country. With his business degree and a Christian worldview, Andrew saw how foreign aid was harming his homeland, particularly in regard to exploitation of his country's coffee bean industry. Uganda is a rich producer of coffee beans but most of them were being shipped out of the country soon after picking. The farmers in his country were not realizing much of the benefit of their hard work, so he decided to change the rules and keep the beans in his country until they were roasted and manufactured. Then the finished product could be sold directly.[12] To help out the farmers, he is sharing profits of his effort with them so that they realize more benefits from their hard work. Together they are hoping to slowly change the economics of their country. They hope to create their own economy. They don't want a government which spends most of its time filling out paperwork to obtain more foreign aid.

As I sat and listened to Andrew's story, I became excited. I thought, "This is how we can really make a difference in the countries around the world." We can actually do good for the people instead of trying to help while actually harming people. This is a real world example that changed how I feel about the work we can do around the world. Andrew found a way to change the world around him. Is there anything we can do to change our world?

I'm not an economist, and I'm not about to start a coffee company. (Though as a person who lived in Seattle for over 20 years I do love my coffee!) I am in the field of technology. Not only as a technician but as an avid user. To me the possibilities in my industry are endless. There are so many different communities online. Within Facebook, Second Life or similar organizations, there are a variety of communities. We all have different talents and interests and, together as the whole body of Christ--The Church--we can make a difference. Through a respectful, loving approach we will learn how we can help not hurt in other countries, nearby states and even in our own cities while we live out our lives online.

***In Summary***

God has been developing these thoughts in my heart for years. It is my hope that just like when I heard Andrew Rugasira and felt energized, that some of you will feel the same when reading this book. I hope you will realize that life online is real and that there are real people with real hurts online. I also hope you realize that these people are very similar to you. They may have bad habits, but so may you and hopefully you can work those out by being online. I hope you choose not to explore your bad habits online while you are being anonymous, but either way I hope you come across somebody who can help. I pray that they will have the talents that will reach out to you at that particular point.

I look forward to hearing from you about the real stories of virtual connections that I know are going to happen. Unless things change dramatically you will be able to find me on Facebook. Hopefully you can also find me connected to iMChurch. I can't wait to hear what God is doing online.

One last thing I want to mention is that, while we get all kinds of different messages about love in the world today, we can find God's definition in 1 Corinthians 13 (called the Love Chapter). It is provided below. It's not always an easy one, but if we as Christians live up to this definition, the world will definitely be a better place.

# 1 CORINTHIANS 13:4-7 ESV

"Love is patient and kind; love does not envy or boast; it is not arrogant or rude. It does not insist on its own way; it is not irritable or resentful; it does not rejoice at wrongdoing, but rejoices with the truth. Love bears all things, believes all things, hopes all things, endures all things."

# About Chester Qualls

I almost feel like I was born in church. I was raised in church and my parents were very regular attendees. We hardly ever missed from what I remember. Even when we were on vacation we went somewhere. I grew up as the over-active boy in the Sunday School class that stressed out my teachers. Most all of my close friends came from the church. After graduating from High School and taking a year to see what the real world was like I decided I wasn't yet ready for the real world and enrolled in college. Without a specific direction for my life, meaning I didn't know what career I wanted (I still don't), I decided on a religion major and after taking Greek I knew that Biblical Studies was the emphasis for me.

College was a great time and I made many new friends including meeting my future wife in my Senior year. She came in as a freshman and I feel in love, she didn't yet feel it but we started out as very good friends.

Because the whole college was a Christian college many of my friends from there were also Christians although over the years I realize that we all have different beliefs and values within the essentials.

Upon graduating with a degree in Biblical Studies and having fallen in love with a beautiful girl I knew that if I wanted to marry her I would have to get a job (people told me so). I didn't want to take my studies further at that moment but I didn't have any other real skills. I approached a distant relative who lived near the college and said I would work for the summer for him for free to be able to put something on my resume. He thought about it and said he would pay me and he would give me a job. I ended up working for him part time for over a year. On my first day in the office he told me that he purchased a computer back a while ago, this was around 1984 so Personal computers were not the norm at the time. I had never played with a computer of any kind, neither had he. So he gave me the manual for the computer which started with a section on Basic Input and Output System or maybe it was redirection pipes. Probably the most boring way to start a computer manual because nobody is interested in that stuff. I got started however and by the end of my time there I had multiple documents and spreadsheets setup for the business operation. But the romance with my girlfriend was moving along well and it was time to find a full time job (I knew, because my boss told me). In my new job I came upon a similar situation, two new computers and an accounting system. Nobody had touched either. I began looking at the accounting system and the requirements and even though it had all been picked by the CPA firm the company used I determined it wouldn't work for us. They agreed. So I had to begin a new search, along with my boss the Controller for the company, for a new accounting program. We found it and decided that we needed to have more computers and a network. We got a company to help setup the network and I worked closely with them so that I could support the system. As part of my learning process one day I was poking around and accidentally blew up the entire server. Actually wiped it completely clean, operating system and all. Stunned that it was possible, and that I had done it, I began re-installing the operating system and once I had that setup and running I restored our backup, luckily we had one. It was very stressful but we recovered and I learned a lot. Up until this point computers were fun but once I reinstalled the network and understood more about how it worked and showed people how to work together on the network I was hooked for life.

Networks are what excite me about technology not so much the hardware. The way people interact, collaborate and contribute to the overall process is the thrill of technology for me. When the Internet became popular I had a slow old fashioned dial up modem and began being connected. I also had customers who got T1 lines for the internet and eventually DSL lines that were cheap enough for everyone but always-on like the expensive T1 lines. We are all now on one big happy network together. Individuals and companies are really just hubs in the grand design. What we have seen happen is amazing but I believe we have just begun to understand the overall impact to us, the way we live, love, work and function day to day has changed. We now carry all this capability around in our pockets (smart phones). As exciting as it is we have also seen unexpected things begin to happen. Some of them allow us to express our darkest sides as well as our bright cheery selves. And this takes me at last back to my beginning. To my faith in God and what I have learned from his Bible. I have been grappling to connect my faith with this passion of mine for a connected world. The possibilities are huge and the hurt is still there. This circle of events is the motivation for this book.

Some additional websites not mentioned in the book and not in the endnotes.

http://www.asocialfaith.com (not mentioned previously but this is the website for this book and hopefully more someday. I hope for us all to share our faith in social ways online and in the real world.)

http://facebook.chesterqualls.com (Me on Facebook)

# Endnotes

[1] David T. Olson, *The American Church in Crisis* (Grand Rapids: Zondervan 2008). In Chapter 2 of this book "The Trajectory of the American Church" Olson provides some analysis of this trend. He shows that the church is not keeping up with the population increases, much less growing. There are of course a few sources of growth in the church at large but overall we are losing the battle for hearts in society today. Read the whole book it's sobering. I believe I have read in other places that this trend is longer than 20 years but don't right now have the source for that.

[2] Often the source for this information regarding divorce comes from the Barna Group. In doing research to support what I have read in other places I came upon this webpage (http://www.barna.org/barna-update/article/5-barna-update/194-born-again-christians-just-as-likely-to-divorce-as-are-non-christians). This is an older study but a newer page also exists not stating the issue in the same way but toning down some of the statements no doubt because George Barna was criticized for pointing out what his research showed. The newer page is here (http://www.barna.org/barna-update/article/15-familykids/42-new-marriage-and-divorce-statistics-released). Apparently the idea that Christians divorce rate was the same as the general population was an unpopular statement among some Christians. Hopefully these links will remain in place but if they are changed you can no doubt find the latest info on the site http://www.barna.org.

[3] http://www.imchurch.com (Website for iMChurch), http://facebook.imchurch.com (iMChurch on Facebook).

[4] A few people who read early releases of this book told me that I shouldn't leave you in the dark about this issue... Do people pray for themselves in the Bible? Or should they? In Psalm 50:15 it says "and call upon me in the day of trouble; I will deliver you, and you shall glorify me." This sounds to me like God is giving us permission. Not convinced? Let's look at Jesus teaching on prayer, In Matthew 6:9 Jesus tells his disciples to "Pray like this", he then starts out with praise to God the Father and after that says "Give us this day our daily bread". Much of the rest of this prayer is somewhat selfish except that we not only ask for forgiveness of our trespasses but are expected to forgive others as well. He follows up the prayer by saying that if you forgive you will be forgiven. Do it! But it looks to me as if it's ok for us to ask about personal matters such as our meals and our mistakes. If you still need more proof then let's look at a man who the Bible claims was loved by God. That would be King David. Many of his prayers are recorded in Psalms. He sung his prayers and many were of a personal nature asking God for his favor. We shouldn't pray only for ourselves but I dare say it's okay to pray for what is on our hearts. So I did a little bit of the work for you but you should still always verify for yourself.

[5] http://www.youversion.com/live - youversion.com is only one of the online Bibles I like to use. Each has different benefits. The live part of youversion.com is pretty slick though. It's a great way to organize the Bible verses you are going to use for a specific event, it allows for various options and I'm sure it will be improved regularly.

[6] Is God really different between the Old Testament (God of wrath) and the New Testament (God of Love)? As I said this provides difficulty for me but I get some clarity when I see God as a parent and we as his children.

Is God really a God of wrath in the Old Testament? In Exodus 34:6,7 he tells Moses " The Lord passed before him and proclaimed, 'The Lord, the Lord, a God merciful and gracious, slow to anger, and abounding in steadfast love and faithfulness, keeping steadfast love for thousands, forgiving iniquity and transgression and sin, but who will by no means clear the guilty, visiting the iniquity of the fathers on the children and the children's children, to the third and the fourth generation.' " There are many other passages in which the compassion of God is mentioned. So is God really different? Or is this merely our perception? I love my children. I would do just about anything for their benefit. That sometimes means that I will let them make a mistake. When they make a mistake it is not my fault for letting them do that, I am being a good parent by letting them learn about the world they live in. My children however do sometimes see me as a mean and vengeful person. They may believe I don't like their friends. They even sometimes believe I have it in for them. That I am out to get them or trap them. That is not true. That is their perception of me. As time goes on my children have come to see that I do love them (hopefully they do). Their lifespan is short compared to mine. Little about me has changed during the duration of their life. I had many experiences before they were born and I still have experiences outside of their existence. So when we see God differently between the Old Testament and New Testament I believe it is something like the experience of a parent. I see God as mean when I have been bad. I see God as vengeful when he helps with my enemies. And I see God as loving when I need it most. He is the parent I am not but I am still me. I am imperfect and the differences I see are not his differences. Does that make sense?

[7] Online harm can come in many forms. You may be tempted visually because of the images available, but can also be wrongly influenced by ideas from others and the discussions you get into. Some of us are strong in one area and weak in another so we can become harmful to each other without meaning to. We can take someone into danger. It may not be danger for us but it is for them. Same for us. We can't go all the places that another can. Unfortunately we often think that we can. That can harm us unsuspectingly. The harms can be emotional, or psychologically damaging, or of course also spiritually damaging if our faith is shaken. We must gently prepare ourselves and others.

[8] http://www.etc4kids.org (Kid's camps my wife is working on)

[9] http://poiema.me (Another site where my wife hopes to help people to see that they are God's Masterpiece.

[10] http://www.twloha.com (website of To Write Love on Her Arms helping people with depression and thoughts of suicide).

[11] lucisix66@gmail.com - Watch out, dangerous email.

[12] http://www.goodafrican.com (website for the coffee company in Uganda)

Made in the USA
Monee, IL
04 November 2026

67375587R00125